LIVING
NOVATION

ıpeting in the 21st Century
Access Economy

LIVING INNOVATION

Competing in the 21st Century Access Economy

HERVÉ MATHE

ESSEC Business School, Paris – Singapore

World Scientific

NEW JERSEY · LONDON · SINGAPORE · BEIJING · SHANGHAI · HONG KONG · TAIPEI · CHENNAI · TOKYO

Published by

World Scientific Publishing Co. Pte. Ltd.
5 Toh Tuck Link, Singapore 596224
USA office: 27 Warren Street, Suite 401-402, Hackensack, NJ 07601
UK office: 57 Shelton Street, Covent Garden, London WC2H 9HE

British Library Cataloguing-in-Publication Data
A catalogue record for this book is available from the British Library.

LIVING INNOVATION
Competing in the 21st Century Access Economy

ISBN 978-981-4719-57-5

In-house Editor: Sandhya Venkatesh

Typeset by Stallion Press
Email: enquiries@stallionpress.com

To my children, Inessa and Lionel, for their patience and understanding

FOREWORD AND ACKNOWLEDGMENTS

The digital revolution has transformed the dynamics of communication and the way of living. The tide of media is giving rise to seamless communication, permanent connection to others and the marketplace, and an expectation for immediate responses, spontaneous purchases, and instant gratification. For the organization, the digital revolution has not only changed the dynamics of communication with the customer, it has also changed the dynamics of innovation. The consumer, voluntarily or not, is increasingly stepping into the innovation laboratories of the organization, traditionally a territory which was closed to "outsiders". The organization, too, is stepping into the living universe of the consumer. Here, increasingly sophisticated market intelligence that pulses with live information from the marketplace can be gathered. Such "live" information is changing the dynamics of innovation which the organizations accomplish through deeper understanding of user needs.

In today's business world where products are much more broadly defined, where the intensity of material continues to decrease as part of GDP, service innovation is proving a key resource for any company or organization. Whatever the industry or government field, in order to succeed, has to focus on the service innovation. In manufacturing environments, service can no longer be viewed merely as support to the production process and a secondary industry in the consumption of goods. Rather, the goal now is to monitor manufacturing systems in order to guarantee they perfectly fit and efficiently respond to the logic of permanent delivery of service value. Fundamental to the service-driven logic are the user-centric perspective of development and the management of the customer relationship, regardless of "product" or conception of offerings in terms of customer "utility". This book precisely addresses numerous issues regarding the consequences of the shift from a product-based economy to an experience-driven logic of value creation. In chapter one, the living innovation paradigm is suggested following a three steps generic process: Live virtual interaction, live face-to-face experimentations, and live deployment. Consequences at organizational level of adopting the living innovation paradigm are then introduced. The rising dominance of the service economy gives matter of development in chapter two with emphasis on how a quest for quality of life may actually drive the shaping of consumer experiences. Three keys to making service innovation a success are illustrated in chapter three with special attention to co-creation and service design. In chapter four, we assessed the challenges of product design and manufacturing reinvention in the access economy; while an integrative approach to product development is shaped in view of practical use in corporate environment. Introduced in chapter five, the concept of imagination ecosystem goes beyond open innovation in the 21st century digitalized business environment where sources of disruption may come from every corner. Organizing for innovation constitutes the topic of chapter six with the functioning of a series of organizational structures which are then analyzed. Next, managing talent for innovation represents the focus of chapter seven. Leveraging the five drivers of the innovation's motivation proves a key objective for companies and developing a culture of innovation a major challenge. Relatively limited attention has been given in the literature to the role of the physical space as a stimulator of innovation. So in chapter eight, we explain how architecture and the design of the working and shopping environments may transform organizations into creative laboratories. Following that line of development, we extend the observation field to look at regional and national environments and their impact to the performance of innovation initiatives. A mapping of the global innovation landscape makes the first part of chapter nine; followed by a discussion on the specifics of the Asian and then Chinese contexts. Finally, attention is given in chapter ten to the critical issue of sustainability and innovation with elaborations on the circular economy and on the reverse innovation process. The bibliography, even if not exhaustive, enlists quantity of materials and may provide some sort of guidance to the reader for further research.

My effort in shaping this book is based on a teaching association with the courses of innovation process management at the ESSEC business school, both in Paris and Singapore, as well as in the numerous institutions I have lectured over the years. My experience as manager and consultant for firms and public institutions, as well as my four-year tenure as Dean of ESSEC Asia–Pacific in Singapore, was a great inspirer of the research conducted to produce this text. The material that I have collected were primarily in case forms as part of the courses taught, and analysis conducted in association with the Institute for Strategy, Service and Innovation (ISSI). This institute that I founded in 2004 has supplied many ideas and illustrations that are commented in the text. As a consequence, I owe a debt of gratitude to leaders of industries, colleagues, and students with whom I have worked and by whom I have been influenced. Among them are Professors James L. Heskett and Roy D. Shapiro, both Emeritus at Harvard Business School, whose inspiring talks have had a strong influence on my development; Professors Daniel Tixier and Xavier Pavie both from ESSEC in France, with whom I have written previous books and conducted many exciting research and educational projects.

While I take responsibility for the key concepts and framework of the material, as well as for any possible inaccuracies, I was aided in this enterprise by two outstanding research associates: Marvyn O'Keeffe, initially with ESSEC in France before joining the industry, contributed to the first versions of the manuscript and Ziying Huang, formerly with ESSEC in Singapore, who has reorganized parts of the materials in the final versions. Ziying Huang has to be credited for the value she added in developing sections on "remanufacturing", and "drivers of innovator's motivation", as well as in producing large parts of the contents of chapters six: "Organizational structures that stimulates innovation" and ten: "Innovating towards sustainability". Working on this book brought me into contact with numerous companies' leaders, public services top administrators, as well as influential consultants and academics from different parts of the world. Many of them have been instrumental in contributing ideas to the material. Finally, I owe debt to Sandhya Venkatesh, representing World Scientific, who has been enthusiastic about publishing it.

CONTENTS

Chapter 1

THE LIVING INNOVATION PARADIGM

"Pearls don't lie on the seashore. If you want one, you must dive for it".

— A Chinese proverb

Technology and the ubiquity of social media provide a lens that enables us to "deep dive" into user environments filled with rich nuanced information. Consumers, who are emboldened and empowered by the freedom of expression and interactive capability, which are facilitated by modern communications, are taking innovation into their own hands. Customers are participating in design processes, be it the co-creation of a customized computer with Dell through the Dell Idea Storm Platform, or the "friendly hacking" of Lego software in order to make a toy robot do the activities that even the company never imagined before.[1] They are influencing the design of new products and services, while shaping the innovation process in an increasingly hazy interface between the organization and the user. As the Chinese proverb portends, it is here that the "pearls" of wisdom which reveals customer needs can be gathered for decryption. In this services-driven economy, where the customer is a co-creator, clients have found their voices more powerful than ever before. The era of the Ford's T Model car, which states that "any color as long as it is black" takes on a mythological quality in today's era of user-centered design and the panoply of choice in the marketplace. Today, Ford capitalizes on its ability to win customers

[1] D Palmer (2007) Harnessing consumer inspiration in new product & service development. *Non-Traditional Techniques for Working with Consumers at the "Fuzzy Front End"*. Walnut Creek, CA: InnovationPoint LLC.

through finely tuned user-customization, which is accomplished *via* an elaborate interactive webpage on which customers can design, dream, and play.

However, both in manufacturing and traditional service businesses, numerous companies are facing unprecedented challenges in the form of a new wave of digital upstarts that not only build their unique value equation on changes in technology, but also on the availability of data to create innovative and customer-friendly alternatives to incumbents' offer. A 2015 McKinsey Quarterly paper emphasized that the sorts of disruptions which materialized in retailing with Amazon or Alibaba, are now invading all industries. Among many others, Uber or Blablacar in mobility, and Airbnb or Tripadvisor in tourism are drastically redefining the new service landscape.[2] Also, the use of materials have dramatically decreased over the past 90 years in terms of the amount of resources extracted per point of GDP. While 280 grams of materials were needed to produce one US dollar of GDP in 1910, only 75 grams were necessary in the year 2000. Among the most typical illustrations of dematerialization, which is frequently mentioned is the production of a typical PC which unveiled a 68% decline in the amount of materials required in 30 years, though each equipment housed 250,000 times more RAM. Even if consumption has significantly risen in absolute terms, organizations use far fewer materials today than they once did to get the same things done owing to the affordability of products and the growth of the global population.[3] Services, entertainment products, and experience-based interactions have become the prime source of economic exchange. Hence to access the full potential, the management of customer experience and of the relationship over an extended period of time has to be addressed in a dramatically redefined economic paradigm.

A Re-Evaluation of the Meaning of the Word "PRODUCT"

While creating the physical product was the point of departure for most companies, now a "product" necessarily entails two more elements:

1. Service experience. It becomes essential to develop a service offer that is underpinned by equipment or any other manufactured items (such

---

[2]T D'Emidio, D Dorton and E Duncan (2015) Service innovation in a digital world. *McKinsey Quarterly*, February. www.mckinsey.com.

[3]V Smil (2015) Vision statement, stuff: When less is more. *Harvard Business Review*, March and *Making the Modern World*: *Materials and Dematerialization*, pp. 36–37.

as technology, which is also serviced). In contrast to the traditional logic of new product development (NPD), the product development is now predominantly driven by the service objective. The product can be sold, rented, remanufactured, refurnished, and resold, thereby plunging it into a "usage-usefulness cycle", a cycle which is optimized by maintaining a vision of its development over time. There exists a renewing capacity (see for instance Apple's continually renewable capacity) as opposed to a "once-off" or infrequent acquisition of a physical product.

2. Customer relationship. As part of the "product management cycle", managing customer relationships effectively and efficiently boosts customer satisfaction and retention rates, and attracts new customers. Customer relationship management explicitly recognizes the long-run value of current and potential customers, and seeks to increase revenues, profits, and shareholder value through developing, maintaining, and enhancing successful company–customer relationships.

Technological innovations have enabled the development of more dynamic approaches to elicit the subtle and often tacit needs of customers. For example, take the usage of social media. Nowadays, social networks are widely used across various industries, whether it is a commercial organization or public service agency: It is heavily used in education (72%), communications (71%), services (66%), and retail/wholesale sectors (61%); while energy and utility companies (41%), manufacturing companies (32%) and government organizations (27%) are less active users.[4] Based on the spectacular development of social networks, numerous creative solutions are emerging in multiple industries, which induce new types of interactions. Today, one of the defining methods of designing and launching new products is asking customers to develop them. Starbucks has deployed a forum for interacting with customers to great success — the Starbucks idea-gathering platform *mystarbucksidea.com.* Launched in 2008, this website makes customers the co-creators of Starbucks' new product development. It allows consumers to submit any idea ranging from the new coffee drink flavor to innovative promotion methods. During its first year, *mystarbucksidea.com* collected 70,000 suggestions. By 2014, the company had implemented over 1,000 changes, and not only launched new services, but most importantly increased customer loyalty

[4] J Davis (2010) The new conversation: Taking social media from talk to action. *Harvard Business Review Analytic Services.* Boston, MA: Harvard Business Publishing.

by showing that the HQ is willing to listen to the voice of the customer (VOC).

The digital revolution has facilitated a paradigm of "Living Innovation" to take precedence: Innovation that capitalizes on the deep understanding of users in their living environments with a real-time relevance to users' lives.

In the context of the fast-paced and protean marketplace, companies can no longer blindly pursue an "ivory tower" approach to new product or service development with their senses closed to the needs of the consumer. Companies not only have a greater degree of connection with their consumers and their ecosystem, but also a greater degree of submersion that is reached through the "living universes", the environments where tests can be made and honest feedback can be obtained. For organizations, the idea of interacting, listening, and observing users in their "living" environments is becoming a key capability. Thus, living labs, social networks, and other methods of interacting and connecting "live" with users are tools that companies can use to capitalize on the knowledge residing in the user environment. Figure 1.1 shows social network usage by industry.

Innovation no longer has to rely on old information and a "hunch"; instead, it can benefit from "live" data that perpetually feeds the innovation machine. In the services economy, tests often have to be conducted in real "living" situations. In the economy of services, the R&D laboratory that develops things in isolated situations has reached its limits. The very

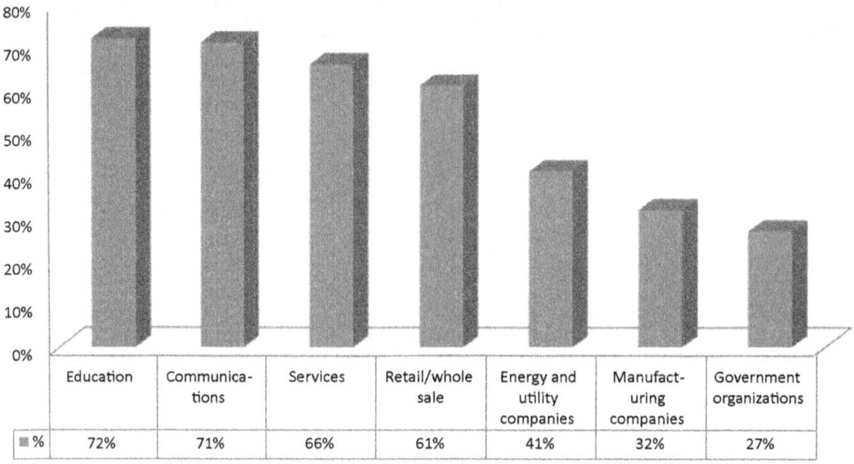

Figure 1.1: Social Network Usage by Industry.

notion of going out into the world and testing in real situations is the new *vademecum* of innovation. The idea of leaving the test lab and reaching out for the real world triggered rapid development of the *tryvertizing* concept, where consumers are granted access to new products, sometimes even concepts and expected to give feedback. A pioneer in this trend was Japan's Sample Lab, a member-only space that allows clients to take five items per visit for an annual fee of 10,000 JPY (84 USD). Company initiatives include Dell, which rolled out kiosks in GameStop stores letting gamers try out new gaming platforms, and Whirlpool's 12,000 square foot *Insperience* Store with fully equipped kitchens where visitors can try out the latest Whirlpool appliances.

Finally, clients have another testing tool in their disposition — the Internet. A virtual world which was launched in 2003 with more than one million active users as of 2014,[5] called Second Life, is a "virtual playground" that enables companies to test their products and estimate the success of upcoming promo campaigns. Around 15% of all Fortune 500 companies did have activities in Second Life, with most of them focusing on marketing and collaboration activities. Although it started as a rather non-priority activity, its presence brought fruitful results with 54% of the visitors giving feedback and 40% receiving publicity messages.[6] Even if less popular in today's world, the Second Life experiment has paved the way for a series of new and inspiring initiatives.

The logic of living innovation follows some of the tenets of both open innovation and design thinking. For the former, the user and other actors in the ecosystem participate in the creativity process either proactively or following an invitation by the company. Instead of relying on the knowledge that lies within the organization, open innovation reaches outside the company to solicit solutions.[7] For design thinking, the process aims to get closer to users, empathize with them, understand their emotions, and glean information for shaping design.

Yet, living innovation is more than involving users in innovation or soliciting their ideas. Living innovation brings the designer closer to the user, on a much deeper scale than was ever possible before. Living

[5]Linden Lab (Developer of Second Life) (2013) Infographic: 10 years of second life. Retrieved 2014. www.lindenlab.com/releases/infographic-10-years-of-second-life.

[6]A Barnetta (2009) Fortune 500 companies in second life. ETH Zurich. Under the supervision of G von Krogh and P Jäger, January. www.smi.ethiz.ch.

[7]HW Chesbrough (2003) *Open Innovation: The New Imperative for Creating and Profiting from Technology*. Boston: Harvard Business School Press.

innovation means bringing live data on users' activities, their "likes" and "dislikes", their social graphs, and their candid reactions to products, services, and brands into the hands of the innovator to signal unarticulated needs. It enables innovations to extend beyond the walls of the organization for testing and even iterating designs firmly in the territory of users where emotions, feedback, and contributions can be gleaned through a myriad of tools. New testing methods look even more appealing in the light of traditional test failures. In 1985, when Coca-Cola Company launched New Coke, it learned about the unreliability of market tests in a hard way since even after unanimous approval of the customers during taste tests, it was forced to reverse the new product three months after its launch because customers did not like the new taste.

The change is happening rapidly and most companies are rushing to arm themselves with the tools and the intelligence to capitalize on this new territory for innovation. While Facebook and Twitter may represent opportunities to gain information from customers and build relationships with clients, it may not be appropriate for every company. Tapping into the goldmine of information is complex. The biggest challenge for companies today is to actually manage the abundance of information by extracting the most pertinent and valuable inputs, and acting upon insights.

In the Living Innovation Paradigm, introduced by Marwyn O'Keeffe and Hervé Mathe in 2011 (see Figure 1.2), the overlap between the user and innovator universes takes place in three key areas: The "Live" Virtual Interaction space, the "Live" Physical Experimentation space, and the "Live" Deployment space. The Virtual Interaction space drives and shapes innovation in the organization. The Face-to-Face Experimentation space is a hazy space that shifts between the user and the organizational universes, depending on the type of experimentation. The Live Deployment space is driven by the organization's service or product offerings. In the user design space, users participate freely in discretionary innovation. Peripheral visioning on the market feeds live data back to the innovator space in the organization.

LIVE VIRTUAL INTERACTION

The usage of social media varies extremely from company to company because most of them are currently searching for the most efficient way to use social media. In a recent Harvard Business Review Analytic Services survey of 2,100 companies, 50% said the main benefit of their activity was

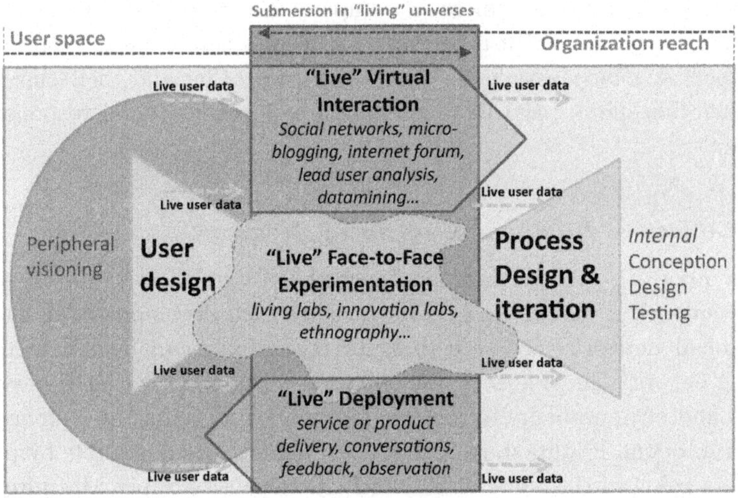

Figure 1.2: Living Innovation Paradigm.

to increase awareness of their organization and its products and services among target customers. 26% said social media usage lead to more favorable perceptions of the organization, products or services. Another 30% saw an increase in traffic to their website as a prime benefit, but only 29% reported collecting and tracking customer reviews on their website or other websites.[8] Thus, a typical US company uses a blog to comment on business issues which are also followed by media reporters, while also using Facebook, LinkedIn, and Twitter to reach customers.

Most of the Web 2.0-savvy companies realize that social media is not about selling products but about developing a community and building a loyal crowd of like-minded people. Consider Timerland's CEO Jeff Swartz, who never uses Twitter to speak about the shoes his company makes but rather about the news from his daily life and social causes he is passionate about. Moreover, in his bio field, rather than the company homepage, he uses the link of Timberland's Earthkeeper project that supports environmental awareness. In this manner, he shows himself as a proactive and responsible citizen rather than just another CEO.

In addition, a balanced multi-channel social media strategy could also be very useful. For example, while establishing the company blog in order

[8]J Davis (2010) The new conversation: Taking social media from talk to action. *Harvard Business Review Analytic Services*. Boston, MA: Harvard Business Publishing.

to share news with customers, a company can also use Twitter to seek and share news. LinkedIn is another platform where company salespeople could post scholarly articles to share with each other and their customers, and their Facebook page to focus on the company's social responsibility efforts.[9]

The Collective Intelligence of Social Network

The number of activities that can benefit from the collective intelligence of social network is astounding. From creative idea development to product test, social networks are here to help. Individuals can benefit from car pooling network Zimride with 300,000 users worldwide[10]; charities can find artists and entrepreneurs to jointly raise money through the HelpersUnite platform; Social Flights make private jet flights more accessible by pairing travelers going in the same direction; French website Super Marmite helps cooks to sell extra portions; Global Points Exchange acts as a peer-to-peer selling platform of frequent flier miles, while Assesscity helps Londoners to identify the simplest routes across the city for those with mobility issues (from pushing a buggy to carrying heavy shopping bags).

For the enterprises, HelpersUnite is a network of small and medium-sized enterprises (SMEs) where firms can buy and sell services like IT assistance, translation or promotion leaflet design; Forrst, a community of developers and designers is focused on sharing their expertise to build better applications and websites; Coroflot is a place for designers to share their work and receive feedback; Naming Force is a platform that crowdsources names and ideas for businesses, websites, and products.

Organizational Zoom . . .

Crowdsourcing Ideas: Eli Lilly and Johnson & Johnson

The term "crowdsourcing" is taken from the words "crowd" and "outsourcing," which was first coined by Jeff Howe in a June 2006 *Wired* magazine article *"The Rise of Crowdsourcing"*.[11]

---

[9] *ibid.*

[10] C Barras (2009) Innovation: How social networking might change the world. *New Scientist.* http://www.newscientist.com/article/dn16681-innovation-how-social-networking-might-change-the-world.html.

[11] J Howe (2006) The rise of crowdsourcing. *Wired* (14), June 6th.

Pharmaceutical maker Eli Lilly funded the launch of InnoCentive in 2001 as a way to tap intelligence from outside his company to help develop pharmaceuticals and increase the speed of their time-to-market. InnoCentive, from the time of its inception, has a transparent policy of allowing other firms to access its network of ad hoc experts. Companies like Boeing, DuPont, and Procter & Gamble (P&G) now post their most ornery scientific problems on InnoCentive's Web site, while anyone on InnoCentive's network can take a shot at cracking them.

Johnson & Johnson, one of the pharmaceutical leaders, delivers "Procrit" an anti-anemia post cancer therapy. www.chemocare.com provides support and assistance to the caregivers of the patients. Launched in 2007, this site is sponsored by Johnson & Johnson. Not foreseen, the company accesses a huge volume of original information such as: Collateral effects, words of mouth impact, new treatment suggestions, etc. Nowadays, the site represents a key source of invaluable data for marketing as well as R&D. For a $50K annual budget, Chemocare.com works and benefits a visiting annual population of 10 million and more.

Analyzing Customer Sentiment in Microblogs

Is there meaning in a "tweet"? Some tweets, or microblog[12] messages, are meaningless chains, a means of self-expression; others are poetic proclamations, or "call to arms". Twitter was commended for its role in facilitating the freedom of expression in Iran during the Arab spring when other media were shut down; it was lambasted for its role in enabling the organization of the riots in London. Whatever role it has, Twitter is a vehicle of candid expression, and its extensive usage across the world is enabling data analysts to gauge the temperature of a nation's mood. The number of tweets sent per day is rising at a colossal rate — over 500 million in 2014, a figure that has doubled since early 2012.[13] Product designers and marketers are particularly interested in the information lying beneath the Twitter phenomenon.

> "*Most tweets are inane, but a million may contain valuable information*".[14]

---

[12]Microblogs are platforms for sending short instant online messages such as Twitter. A tweet is a message sent from the microblog Twitter.

[13]R Krikorian (VP, Platform Engineering, Twitter Inc.) (2013) New Tweets per second records, and how! Twitter Official Blog. https://blog.twitter.com/2013/new-tweets-per-second-record-and-how.

[14]The Economist (2011) *A market for social-media data: Sipping from the fire hose.* October 1st. http://www.economist.com/node/21531025.

Companies such as Datasift and Gnip are amongst the first to launch a data platform where they collect and categorize information from all kinds of social-media services: Twitter, Facebook, YouTube, and others. These companies charge a monthly fee for a feed of tweets, which are then purchased by companies such as Sysomos. The goldmine of information lies in the analysis of these feeds. Without a meaningful analysis or visualization of the data, it is effectively useless. Sysomos allows companies to track in real time what people think about certain products. This represents a real value in terms of product or design. Tests can be made in real environments and the data can be tracked to measure customer sentiment through their reactions to data. Lexalytics analyzes the sentiments expressed in these messages and posts. Klout measures the influence of social media users.[15] Accessing live data on consumers comprises certain challenges, e.g., the data is not cheap. Gnip, an American company, charges $33,000 per month for a feed of half of all tweets. Service at this price only provides raw data, and further analysis to refine and map the data represents a further cost.[16]

Lead User Analysis

Involving customers, carrying out market research and questioning needs, while often extremely valuable, is also considered to be somewhat outdated. To create or anticipate markets for truly new and innovative products, what Eric Von Hippel refers to as "lead users" are instrumental in the type of market research that is required for today's markets. In fact, customers whose expectations go far beyond those of regular customers can offer richer insights because they are the ones for whom the product performance matters the most.

Organizational Zoom . . .

Sense Worldwide: Innovative Ideas from Extreme Users

Sense Worldwide, a London-based innovation consulting firm, is among an increasing number of consultancies and design teams that specialize in working with extreme customers. It has been featured in Wired

[15] *ibid.*
[16] *ibid.*

> Magazine[17] and Fast Company[18] for using lead users such as Nigerian hackers, medical tourists, teddy-bear enthusiasts, and obsessive compulsives in their innovation projects.
>
> For example, in a project to help a client develop a toilet brush, by observing customers with an obsessive compulsion for cleaning their lavatories hygienically, the research team from Sense Worldwide found that many of them wrapped the bristles in toilet paper to minimize contamination. This observation gave Sense Worldwide the idea of flush-away biodegradable covers for the brush-heads. "Since [the participants] were so interested in the minutiae of cleaning, they were willing to go into details that most people would rather not think about," says Brian Millar, Director of strategy at Sense Worldwide.[19]

In high technology industries, new product development occurs so rapidly that the role of "lead users" assumes new significance. The "real-world" experience of the ordinary user is often rendered obsolete by the time a product is developed or during the time of its projected commercial lifetime. Eric Von Hippel defines a lead user as the one whose needs reflect the needs of the general market, only months or years before the mass market; and who stands to benefit from the solutions to his needs. Von Hippel argues that lead users provide key insights to future market needs and can therefore play a constructive role in the design of truly innovative products or services. Many of the products that we seek today are no longer need-based, but desire-based. Indeed, as Tim Brown, the CEO and President of IDEO writes, "*As more of our basic needs are met, we increasingly expect sophisticated experiences that are emotionally satisfying and meaningful*".[20] GE calls its lead users as "luminaries". According to Sholom Ackelsberg of GE Healthcare, GE then shares some of its advanced technology with a selected subset of these "luminaries" who form an "inner sanctum of good friends", a collaboration that gives rise to GE products.

---

[17]Wired (2012) Need extreme ideas? Talk to extreme consumers. http://www.wired.co.uk/news/archive/2012-04/30/extreme-consumers.

[18]Fast Company (2012) Want breakthrough ideas? First, listen to the freaks and geeks. http://www.fastcodesign.com/1669452/want-breakthrough-ideas-first-listen-to-the-freaks-and-geeks.

[19]A Clegg (2014) Harnessing the power of extreme consumers. http://www.ft.com/intl/cms/s/0/f7256696-746e-11e3-9125-00144feabdc0.html#axzz3OV3UnQ5S.

[20]T Brown (2008) Design thinking. *Harvard Business Review* 86(6), p. 84, June 1st.

"LIVE" FACE-TO-FACE EXPERIMENTATIONS

Chad Duhon, the Director of product marketing for the IdeaPad brand of Lenovo, the Chinese PC maker acquired from the PC branch of IBM, argues that the ultimate "sanity check" for ensuring that a product appeals to users is to confront your product or service with users. As Duhon remarks, *"you've got to show your product to people who have nothing to lose by giving you an honest opinion and will tell you if your baby is ugly".*[21] Especially in today's market, getting closer to the consumer is one of the principal by-products of the technological revolution, but it should not be limited to websites, apps, and social networks that have irrevocably rewritten the codes of communication between the company and the user.

Living Labs

Living labs are open innovation platforms for user driven multi-stakeholder open innovation involving the citizen and the user in a laboratory context. The experimental context offers companies the perfect playground to experiment with various approaches in a risk-free environment. Smart, innovative companies are getting engaged with the consumer through such "living laboratories" where they can dialogue with users in an experimental setting. Orange has embraced the "living laboratory" approach with its experimentation lab that contains immersive infrastructure for 3D experiments, a platform of mobile devices for simulated environment (Ubiquarium) and a fully instrumented apartment, which enabled the company to acutely understand the importance of tests in semi-real environments. Transportation, banks, and other sectors are embarking on similar trends.

Laboratory tests are not necessarily reserved for corporations that have large funds to develop them. With the recent development of FabLabs[22] in Manchester by the Manufacturing Institute and its 20 branches worldwide, every single person ranging from a small child to an entrepreneur can have access to fully equipped lab with 3D printers, milling, sewing, and laser cutter machines, and can thus transform their concepts into real product prototypes.

---

[21] J Bird (2011) Customer research: Testing the market is a vital 'sanity check'. *Financial Times*, October 18th.

[22] Fab Lab company website http://www.fablabmanchester.org/.

Organizational Zoom ...

Social Networks as Extended Living Labs

Researchers at Microsoft have developed a game on Facebook that functions as a virtual laboratory to study how people play strategic games. Rather than bringing a group of people into a physical laboratory to test game theory, Facebook instead can be used as a giant, global laboratory, which enables them to massively increase the scope for testing; previously limited to 35 participants in a physical lab.[23] The game was designed at The Microsoft Research Lab at Cambridge University, a lab that is known for its success through its Xbox Kinect and Artificial Intelligence technology. The artificial setting of the traditional lab is replaced by the user's circle of friends, their "natural social habitat",[24] referred to as the "social graph" by Facebook CEO Mark Zuckerberg.[25] People are becoming increasingly comfortable with the interactions on the virtual landscape. Furthermore, in some cases, the virtual environment of the social graph is a more accurate reflection of reality and sincere customer response than the semi-real environments which are devised to test things live with users.

Empathy and Immersion

Innovation often starts with an insight about a latent or an overlooked need, and such insights are most often prompted through empathizing with your customers in real "live" situations. Companies are embracing the realization that users are real people, and thus gaining a deeper understanding about what such people experience throughout their daily lives can reveal insights that spark fresh thinking and enrich the innovation process. Consider the example where rail authorities in London realized that a negative experience was created as commuters waited for Metro trains to arrive at the station. A deeper discovery revealed that the emotional anxiety wasn't due to the wait *per se*. Instead, it was due to uncertainty about the arrival of the next

[23]Forbes (2011) Microsoft uses Facebook as giant 'lab' to study game theory. http://www.forbes.com/sites/parmyolson/2011/10/12/microsoft-uses-facebook-as-giant-lab-to-study-game-theory/.

[24]M Foley (2011) Microsoft readies a 'Facebook game theory lab', all about Microsoft blog. http://www.zdnet.com/blog/microsoft/microsoft-readies-a-facebook-game-theory-lab/10803. Microsoft Research: http://research.microsoft.com/en-us/projects/research games/.

[25]The "social graph", refers to the "people and the connections they have to everything they care about". https://developers.facebook.com/docs/reference/api/.

train. Once officials installed digital displays indicating the arrival time of the next train, customer experience statistics improved. In retrospect, it is easy to see how replacing ambiguity with order reduces anxiety, but it took decades of poor customer experience before an emotional insight was leveraged to design a dramatically improved customer experience.

> *Empathy is not walking in another's shoes. First, you must remove your own.*
>
> — Scott Cook, Founder, Intuit

In the fashion industry, companies are also realizing the importance of empathy in their innovative development. Damart, a French fashion brand for older female customers, took this approach for designing clothes that would appeal to the target customers. The company undertook a process of going out into the world and discovering who the customer was, what made her tick, and what constituted an average day for her. The research revealed insights that enabled the company to develop personalities, names, and preferences for a series of "model" customers and to design better clothes based on this understanding. Velvet Brigade, San Francisco-based fashion startup pushed this concept even further by submitting the drawings of aspiring designers for Twitter voting and then manufacturing the models selected by the public.[26] In this case, the company was literally fueled by the public because the ideas were selected and the decisions were made by the clients, not the managers of the company.

Immersion into the living universes of the client often ignites creative thinking and problem-solving abilities more effectively than aiming to solve problems for the client remotely. IBM has found an effective strategy for customer-oriented innovation through its *First-of-a-kind* program. This program entails sending IBM research personnel directly onto the site of an IBM customer who is tackling a tricky problem. The immersion in the living environment of the client enables IBM to test pioneering technologies on real live business problems. Since its launch in 1995, the program has led to over 150 innovation projects developed through problem solving face-to-face with the client. Examples of successful solutions and innovations included providing key information for healthcare monitoring without compromising patient privacy, developing a real-time warning system for troops in Iraq,

---

[26]http://www.springwise.com/fashion_beauty/fashion-label-crowdsourcing-design-product-range/.

and driving down the cost of electricity by incorporating smart systems in power grids.[27]

Organizational Zoom . . .

IBM Reconnects with the Customer: Empathy and Intimate Understanding of the Customer to Redesign the Business[28]

IBM faced a major crisis in the early 1990s — the company strategy and the erstwhile business model based on mainframe computers was no longer relevant in the market. The company was rapidly losing money and was letting go thousands of employees — the giant was facing the biggest crisis in its history. A new CEO, Lou Gerstner, was brought in to resolve the crisis and the expectation was that the company would be broken down into a series of smaller companies. But instead, Gerstner sought a way to enable the company to survive. To do this, he launched "Operation Bear Hug"[29] — an empathy program designed to listen and discover more about the customer — he sent his top 250 managers out into the world, each to visit at least five customers in person, to listen to their concerns and their thinking on how IBM might be able to help. The empathic connection to real-world customers achieved through Bear Hug helped managers to see whether the decision added value to customers or destroyed it.[30]

The results were instantaneous — it immediately led to more reactive resolution of customer problems, as well as greater attention to new market opportunities. Managers discovered that large corporate clients were fascinated by the Internet, but were not clear about what they should do about it. IBM realized that it could make a major impact by providing the infrastructure needed to help large enterprises leverage the power of the Web. The resulting e-business initiative was wildly successful and helped put IBM on the path to long-term growth. Over time, the company became more and more customer oriented and by Gerstner's second year at IBM, the company had become profitable again, setting off a decade of consistent double-digit revenue and earnings growth. Empathy, which was the cornerstone of design thinking, had

[27] IBM First-of-a-Kind Program (FOAK): http://www.research.ibm.com/FOAK/.
[28] D Patnaik (2009) Innovation starts with empathy. *Design Mind.*
[29] D Patnaik (2010) The Fundamentals of innovation. *Business Week.* http://www.businessweek.com/innovate/content/feb2010/id2010028_823268.htm.
[30] D Patnaik (2009) Innovation starts with empathy: The importance of developing deep connections with the people you serve. *Financial Times.* http://www.ftpress.com/articles/article.aspx?p=1399911&seqNum=4.

changed the outlook of the company to become distinctly customer-oriented and capable of anticipating the complex needs of customers.

Customer Observation in Living Environments: Ethnography

Observation of users in their living environments can often reveal discrepancies between the engineer's designs that often develop in an isolated development laboratory and the user's experience of putting these designs into practice. Such discrepancies can today be easily remedied through the use of live surveillance of user experience *via* postings on social networks, feedback given to frontline staff, or in simulated environments such as "living labs" or innovation labs. A good example is the redevelopment of the Xerox copy machine which was re-engineered to ensure it was more "user friendly".

Organizational Zoom ...

Xerox: Make Device Better by Studying What People Actually Do

The design of the Xerox office photocopy machine, as we know it today, was greatly influenced by user insights, which anthropologist Lucy Suchman brought to the attention of designers at the famous Xerox Palo Alto Research Center (PARC), a subsidiary of Xerox. Field study methods and video technology were utilized to create a film of office workers struggling to operate their Xerox machine, which was then used to sensitize engineers to produce a more user-friendly design.[31] PARC-pioneered ethnography has proved very useful in identifying unmet needs and recommends practices that work, and create products and services that sell.

Recently, Xerox has also applied the ethnography approach in helping the local government in Los Angeles to design algorithms and implement dynamic parking systems.[32] Police meter monitors are now fed data about parking; yet reacting in a way that is rational which leads to sub-optimal results, such as discovering that a car has left the spot in question. David Cummins, SVP, Parking and Justice Solutions said "Ethnography helps in real time by showing that people may be having

[31] E Sanders (2002) Special section: Ethnography in NPD research — How "applied ethnography" can improve your NPD research process. *Visions*, April. http://www.pdma.org/visions/apr02/applied.html.
[32] Forbes (2012) How Xerox uses analytics, big data and ethnography to help government solve "big problem". http://www.forbes.com/sites/benkerschberg/2012/10/22/how-xerox-uses-analytics-big-data-and-ethnography-to-help-government-solve-big-problems/.

difficulties with parking meters. When you have ethnographers out there taking actual videotape anecdotes to show to public officials that people are indeed having a difficult time understanding parking signage, it resonates much more strongly than simply telling them".

Observations can also reveal important insights that can be used to develop more market-oriented designs. Over the course of the past several decades, new product and service development has evolved with the growing realization that the observation of people in real environments is different from asking them what they want. This awareness has led to the adaptation of a range of social science techniques for use in product development. The practice of user-oriented design is not new. The process of "design for usability" emerged in response to the insight that the design of manufacturing equipment revealed significant failures without a sufficiently deep understanding of intended users' skills.[33] Many companies run their own internal usability laboratories to ensure that the final prototype ascribes to the consumers' perception of usability.

While ethnography has been used in the social sciences for a long time, its use in business is a more recent occurrence. Ethnography has been defined as "*a descriptive, qualitative market research methodology for studying the customer in relation to his or her environment. Researchers spend time in the field observing customers and their environment to acquire a deep understanding of customers' lifestyles or cultures as a basis for understanding their needs and problems better*".[34]

Ethnography can be employed at many stages of the design process and in many customer settings (home, office, hospital, bank, and factory etc.) to identify more pertinent insights in relation to the use and experience of the product or service, and it can equally be used in the conception stages to help identify tacit customer needs. The goal is to make implicit needs explicit by describing how customers and users experience products in practice rather than how they report these experiences in a traditional marketing research settings. The use of in-home visits is well accepted for consumer market analysis and segmentation.[35]

[33] P Adler and T Winograd (eds.) (1992) *Usability: Turning Technologies into Tools.* New York: Oxford University Press.

[34] P Belliveau, A Griffin and S Somermeyer (2002) *The PDMA Toolbook for New Product Development.* New York: Wiley and Sons, p. 442.

[35] R Davis (1996) Market analysis and segmentation issues for new consumer products. In *The PDM Handbook of New Product Development.* M Rosenau, Jr. (ed.). New York: Wiley.

Organizational Zoom . . .

Starbucks: Leveraging Ethnography for Enhanced Experience

Howard Schultz famously developed the Starbucks coffee house concept after first observing goings-on in Italian coffee houses during a visit to Milan. While the company also relies on its frontline barista staff to share customer insights, the company officially started carrying out ethnography in 2002. They started taking teams from product development and other company departments on "inspirational" field trips to observe customers and trends. Michelle Gass, the company's senior Vice-President for category management, took her team to Paris, Düsseldorf, and London to visit Starbucks in these very different localities, in addition to the visits to other restaurants to get a better sense of local cultures, behaviors, and fashions. Gass noted that "*you come back just full of different ideas and different ways to think about things than you would, had you read about it in a magazine or e-mail*".[36]

3M: Observing Users in Real Environments[37]

Around 2003, 3M began noticing that customers were using laptops, cell phones, and BlackBerrys to send quick memos or jot down bits of information, which meant using less Post-it notes. However, while Post-it note use was decreasing, 3M spotted a gap in the market for a new product. While observing consumers, 3M researchers asked to see their photos but found that it was not a straightforward process because consumers would scroll through screen upon screen of photos or had to dig through a drawer for the few shots they printed. Nine months later, the 3M design team came up with the idea of Post-it picture paper, or photo paper coated with adhesive that allowed people stick their photos to a wall for display. As Jack Truong, the company's Vice-President of the office supply division commented, "We listened carefully to what consumers didn't say and observed what they did."

Co-Creation

Co-creation becomes the key to improving customer experience in general. Just like Starbucks asked its customers what they want to see changed,

[36] BusinessWeek (2006) The world's most innovative companies. http://www.businessweek.com/magazine/content/06_17/b3981401.htm.
[37] BusinessWeek (2006) The world's most innovative companies. http://www.businessweek.com/stories/2006-04-23/the-worlds-most-innovative-companies.

in the same way the Swedish Public Service reached for its citizens to understand why they were using public transport less often. The intensive communication with clients showed that the problem consisted not in the transportation service itself, but rather in the outside environment where the service was consumed, i.e., people were concerned with traffic congestion and increased pollution. On receiving customers' input, the Swedish Public Service officers developed longer routes, thus decreasing the number of the buses on the road and implemented new IT system to streamline the activities for developing the most efficient schedule.[38]

Companies have broad array of tools to co-create with the customers. "Real-time visuals" invite the consumers to share their issues, concerns and product ideas while an artist captures them in real time; "Here's what I *really* want" technique is a sincere face-to-face talk of customers with a software engineer, a retailer, or the head of product design about their issues, frustrations, and ideas while the designer draws down their ideas in real time; "Devotees and detractors" is the moderated discussion of two "debate teams" of consumers on the (de-)merits of competing brands, or the older *versus* the latest version of a product.[39] Idea contest is another highly effective tool which is massively implemented in the universe of consumer goods. Staples, an office supplies and stationary store held an "Invention Quest" competition among consumers to come up with new product ideas and received 8,300 submissions. The winning idea was launched in 2005 in the form of a gadget called a "Wordlock", a padlock that uses words instead of numbers. IKEA conducted an "ingenious people" contest that asked amateurs to submit designs for storing home media (hifi sets, TV, and DVD etc.) in the living room. Out of 5,000 ideas submitted, 14 were invited to IKEA headquarters in order to attend a workshop and receive a cash prize.

LIVE DEPLOYMENT

Two major trends have changed the face of customer relationships today, which in turn have a bearing on a company's innovations. One is the rise of the service-dominant logic that has impacted firm-level innovation. Service

[38]A Lundkvist and A Yakhlef (2004) Customer involvement in new service development: A conversational approach. *Anders, Managing Service Quality* 14(2/3).

[39]D Palmer (2007) Harnessing consumer inspiration in new product & service development. *Non-Traditional Techniques for Working with Consumers at the "Fuzzy Front End"*. Walnut Creek, CA: Managing Principal, InnovationPoint LLC.

economy means that the division between the production, innovation, and consumption processes is increasingly blurred. Purchase of goods or services has become a value chain in which the customer is intimately involved. The days of a clear-cut transaction when contact with the purchaser occurred solely at the site of purchase are past and customer interaction often begins in advance of purchase and is followed up with after-sales service. Indeed, Mathe and Shapiro (1993) noted that after-sales and delivery services would play an increasingly crucial role in the success of manufacturing firms.[40]

Organizational Zoom . . .

Zappos: Experience is an Investment

Zappos, the frequent leader in client satisfaction ranking had acquired its loyal customer base, owing to mainly the exceeding delivery expectations of its clients.

Instead of spending much money on traditional media, Zappos invested back into its customers and looked to improve customers' experience. For example, the company offers free shipping in two ways, a 365-day return policy and a top-notch customer service. Moreover, even though their website states that delivery will take between 2–5 business days, Zappos gives "surprise" free upgrades to overnight shipping for most customers.[41]

Its CEO Tony Hsieh commented "Our business is based on repeat customers and word of mouth. There's a lot of value in building up our brand name and what it stands for. We view the money that we spend on customer service as marketing money that improves our brand".[42]

Another trend is rise of social media. The increasing use of social networks and internet forums by the customers have created a generation of "empowered users" whereby the customers can tweet, post or comment "live" to entire communities on the merits or shortcomings of a product or service. Small enterprises can afford to put social media in the core of their business model, while bigger companies find it an excellent source of client feedback.

[40] H Mathe and RS Shapiro (1993) *Integrating Service Strategy in the Manufacturing Company*. London: Chapman & Hall.

[41] K McFarland (2008) Why Zappos offers new hires $2,000 to quit. *BusinessWeek*. http://www.businessweek.com/smallbiz/content/sep2008/sb20080916_288698.htm.

[42] Forbes (2014) Five customer-centric marketing lessons from Apple to Zappos. http://www.forbes.com/sites/stanphelps/2014/08/18/five-customer-centric-marketing-lessons-from-apple-to-zappos/.

According to a recent study by MarketTools (MetrixLab), most companies are still using traditional channels for grabbing information with 94% of the respondents not using Facebook, Twitter or any other social media tools. These results constitute the evidence of "feedback myopia", with 92% of respondents acknowledging the importance of keeping customers satisfied, but with only 43% actively seeking customer feedback through traditional sources like email/online surveys (51%), formal phone surveys (28%), and informal phone calls (28%) and with 6% using the most easily accessible feedback tool — the social media.[43] Still, the companies that use social media are modest about their progress: Only a small group (12%) of the companies in the survey said they felt they were currently effective users of social media. The explanation to such situation has to do with mentality of the managers and acceptance of the new tools. According to Avinash Kaushik, Google's Analytics Evangelist, "Too many companies have not evolved from what I call 'shout marketing' — TV, newspapers, and magazine ads that influence by initiating and participating in conversations with consumers". He said "There needs to be a generational shift".[44] Among those 6% of the companies that are active in the Web 2.0 Universe, the majority are large corporations who cannot afford paying the price of bad publicity and unanswered client complaints. Earlier, marketers were trained to believe that a dissatisfied customer tells his woes to ten others. "But in the new age of social media," Paul Gilin, author of *The New Influencers* says, "he or she has the tools to tell ten million."

On the brighter side, many major corporations like Southwest Airlines and Comcast have special teams which are dedicated to monitoring mentions of the company on Twitter and Facebook, making sure that customers' concerns are addressed.[45] Currently the situation is changing for better, with nearly 25% providing customer support through Twitter. Some go even further, such as Whole Foods Market, which gathered advice from their Twitter followers concerning the future product selections and the Home Depot asking customers' opinions about future deals, and

---

[43] E Cachette (2011) 94% of companies do not use social media for user feedback. *SocialTimes.* http://socialtimes.com/94-of-companies-do-not-use-social-media-for-user-feedback_b35025.

[44] J Davis (2010) The new conversation: Taking social media from talk to action. *Harvard Business Review Analytic Services.* Boston, MA: Harvard Business Publishing.

[45] S Gordhamer (2009) 4 ways social media is changing business. September 22. http://mashable.com/.

subsequently modifying them.[46] To cite examples of SMEs, we can mention about the Irish Tweet Café[47] which accepts orders only through Twitter, or the US-based Kogi Korean BBQ, a taco truck whose location can be identified only by following its Twitter account.[48] Finally, social media is helping in public service as showed by BuitenBeter app,[49] which allows Netherlands citizens to submit their complaints and suggestions directly to the municipality about the things or issues to be fixed in the city.

This live and proactive conversation comes rife with risks because such forums lie outside the control of the company. Rather than hiding from the "problem", smart companies increasingly engage in this ongoing dialogue and use it to leverage the elusive customer approval, fish for ideas for improvements in products or services, or provide *de facto* after-sales service. The "empowered" user or customer has greater access to knowledge about products and services and can also act upon this knowledge by publicly expressing opinions in the open forum of the internet. Users and customers are increasingly informed and empowered in their purchasing decisions due to greater access to information *via* the internet and can thus evaluate products, compare prices, and shop around. The service or product provider is thus in a situation of greater transparency *vis-à-vis* the competition and is compelled to design products or services that are more user-oriented.

LIVING INNOVATION IN THE ORGANIZATION

Innovation can be a slow process. People in the organization become consumed with the day-to-day tasks and finding time to innovate can be a challenge. Innovation is often hampered by the size of the organization, the difficulty of getting all the actors together to enable a rapid and efficient prototyping and design. In the avant-garde, we can see that the companies have started to adopt the design thinking approach when developing new products. They turn to focus on what people want and need in their lives and what they like or dislike about the way particular products are made, packaged, marketed, sold, and supported.

An ability to think as a designer requires first of all empathy and ability to put yourself in your customer's shoes. This quality was in the

[46]http://www.searchenginejournal.com/16-examples-of-huge-brands-using-twitter-for-business/7792/.
[47]http://www.springwise.com/food_beverage/pop-up-cafe-takes-orders-twitter/.
[48]http://www.latimes.com/features/la-fo-kogi11-2009feb11,0,4771256.story.
[49]http://www.springwise.com/telecom_mobile/buitenbeter/.

heart of Shimano's makeover, a Japanese manufacturer of bicycle parts that was concerned by the slow growth of racing and mountain bikes market in the US. Being determined to find out the reasons of this decline, the company launched massive market research and found out that for 90% of the adults cycling is associated with childhood memories, while the factors that keep them from cycling in adulthood are security doubts, the challenge of retail experience, and the complexity of maintenance. Having understood how customer thinks, the company developed a simple bike with a comfortable seat, that required little maintenance and was easy to operate. With a simplified retail experience and promotion campaign that indicated the places safe for riding, Shimano tapped into the unexplored market and managed to reach the customers who did not consider cycling as a transportation option before.[50]

While most managers think that having innovation lab is a costly project, in reality it can be developed in a few days with the use of simple tools like furniture on wheels which creates a fluid work space that adapts to the needs of the participants; and large, mobile whiteboards for idea-sharing and presentation stage facilitate the exchange of complex ideas.[51] Finally, it is worth saying that having innovative test spaces and creative laboratories is equally important for service organizations as it is to manufacturing companies. A branch of major advertising agency Ogilvy Digital has launched Innovation Lab Launch in July 2011. This laboratory is an incubator that showcases and helps to develop and market digital innovations. It is a showroom full of all new technologies used for creation of new applications, tools, and utilities and can also be used as an educational facility both for employees and partners.[52]

---

[50]T Brown (2008) Design thinking. *Harvard Business Review* 86(6), p. 84, June 1st.

[51]Innovation Labs. Innovation labs create game changing innovation quickly! http://www.innovationlabs.com/innovation-services/innovation-labs/.

[52]Ogilvy Digital Innovation Lab Launch. http://www.jonathanmacdonald.com/?p=1300.

Chapter 2

SHIFTING FROM PRODUCT TO EXPERIENCE

Today, numerous companies are re-examining the source of their distinctive advantage in the face of uncertainties, continuous flow of emerging technologies and new consumption patterns. The necessity, to invest in sustainable development, to accommodate an exploding world population, to deal with the growing scarcity of raw materials and energy constraints, is leading to a reversal of the traditional logic of value creation. As a result, economies are turning their focus toward demand-driven production of utilities rather than goods alone. New strategies are deployed to utilize technologies that stimulate innovation in services and encourage partnerships. Services, when combined with digital transformation, constitute the very heart of responsible value creation in the 21st century global economy.

The rules and criteria for success have changed in today's market; we no longer find ourselves competing in the traditional economics of demand, supply, and price competition of the commodity market. Earlier, the market was dominated by simple and standalone manufactured goods, however, a hybridized product–service model dominates across all sectors today. A physical product is no longer a sufficient basis for fruitful exchange. Instead, a more complex service becomes the selling point, where the so-called product acts merely as a platform to support the value-producing service. Essentially, the manufacturing component of a business now plays the support role in the indispensable *service logic* of today's markets. The consumption of services is growing in importance, whatever may be the considered industry, country, or social context. Service-focused innovations may actually reinvent entire sectors of the economy, while significantly altering the codes of communication and relationships in our societies.

Innovative services strategies materialize in terms of new and redefined service products, improved client interfaces, new delivery processes, and redefined organizational structures. Above all, the consumer is fueling what is now called the *Experience Economy*.

THE RISING DOMINANCE OF THE SERVICES ECONOMY

While the manufacturing industry was once regarded as a dominant model in economic value creation, today it is no longer the case. For the developed countries, the value derived from increases in manufacturing productivity has reached its limits, and no longer accounts for the principle source of value in these countries. Western economies continuously add positions in services but shed jobs in manufacturing. Indeed, the share of manufacturing as a source of employment has been falling continually in all developed nations over the past several decades. Fewer than 10% of Americans are employed in manufacturing now, while this figure stood at 25% in 1970.[1] Similarly, in Britain, the share of manufacturing as a percentage of total employment has undergone the sharpest decline worldwide, i.e., manufacturing jobs have dropped from 35% to 14% of the workforce.[2] Today, no developed economy not even economies that have a relatively high level of manufacturing such as Germany or Japan, generate jobs in manufacturing. In the developing countries such as China, Bangladesh, and Cambodia, superior labor efficiency techniques are widely in use. In the long run, gains and innovations in manufacturing bring even higher levels of automation that evidently occurs at the expense of jobs. Between 1995 and 2002, nearly 22 million manufacturing jobs disappeared from the global economy despite numerous policy efforts to promote employment in that sector.[3]

Today, both developed and developing economies face the challenge of increasing the labor-intensive services sector to compensate for the decline in labor in the manufacturing industries where jobs are being obliterated by automation, new technologies, and more efficient use of labor.[4] The expansion of services stands as the most spectacular business trend. In

[1] The Economist (2005) *Industrial metamorphosis*. September 29th.
[2] *ibid.*
[3] R Dobbs and R Villinger (2010) South Korea: Finding its place on the world stage: Beyond manufacturing. *McKinsey Quarterly*, April.
[4] *ibid.*

the largest developed economies such as America, Britain, and Australia, services account for some 75% of total national income employing a massive 80% of the total available labor force.[5] In addition, when compared to manufacturing, the service sector requires a greater proportion of high-skilled labor. In the United States, for instance, at least 30% of service jobs are categorized as high-skill (managerial, professional, technical or administrative) compared with merely 12% requiring high-skill labor in manufacturing. Rising incomes are commensurate with the increase in skill levels that in turn lead to superior quality of living standards in a country. Instead of being at the bottom of the standard of living pyramid, families find that the supplementary income gives them access to services that can enhance the quality of life such as entertainment, better health care, telecommunications, travel, and even just a trip to the beauty parlor.[6]

OUTTHINKING THE COMMODITIZATION TRAP

> "Many product companies are in the business of delivering services; they just haven't realized it yet".
>
> — Reinartz and Wolfgang Ulaga (2008)

A Utility and Service-Driven Logic

Companies, even those involved in pure manufacturing, thinking exclusively in terms of a narrow product-driven logic, are at great risk of limiting the scope of potential value creation and future development. Peter Drucker appropriately stated *"What the customer buys and considers value is never the product. It is always a utility, that is, what a product does for him"*.[7] Drucker's considerably broader view of the supply and demand dynamics facilitates a more comprehensive and satisfying product or service development. Similarly, Theodore Levitt once famously told his M.B.A. students at Harvard: *"People don't want to buy a quarter-inch drill. They want a quarter-inch hole"*. Levitt hit the nail on the head, when he argued that consumers are more focused on the experience and outcome of a particular commodity or service they buy, rather than the mere physical presence of

[5]The Economist (2006) *Not yet free to serve*. February 16th.
[6]R Dobbs and R Villinger (2010) South Korea: Finding its place on the world stage: Beyond manufacturing. *McKinsey Quarterly*, April.
[7]P Drucker (1974) *Management: Tasks, Responsibilities, Practices*. New York: Harper & Row.

the product or service. With the commoditization of products, achieving competitive advantage merely by the physical characteristics of a product is outdated. Instead, innovations in providing optimum service levels that the consumer desires will translate into infinite opportunities for companies in establishing a sustainable relationship with the consumers. Indeed, today's service-dominant market has come a long way since the Ford Model T, *"any color as long as it's black"*. Products are increasingly differentiated by the added value they deliver through the service component of the offering.

Increasingly, and appropriately, manufacturers are competing to enhance the value proposition of the product by offering ancillary services, thus creating a more durable and valued relationship with the customer. The term "*servitization*" was first coined by Vandermerwe and Rada (1988)[8] to describe the tendency of manufacturing firms to "offer fuller market package or bundles of customer-focused combinations of goods, services, support, self-service, and knowledge". They pointed out that "services are essentially intangible and are performed, not produced".

One of the outstanding advantages to be leveraged through the "*servitization*"[9] of a product is the creation of a relationship that the customer values to the extent that an exit barrier is essentially established. In fact, services can be a powerful way to lock-in customers and increase their switch-over costs. As one manager at Air Liquide, a leading French company supplying industrial gases and services to various industries including medical, chemical, and electronic manufacturers, expressed, "*the more we enter into a customer's business, the more the customer forgets how things are done*".[10]

Organizational Zoom ...

Xerox: Services to Survive[11]

Over the decades, Xerox's transition from a product-oriented to a service-oriented company has been a long and difficult journey. Far from

[8]S Vandermerwe and J Rada (1988) Servitization of business: Adding value by adding services. *European Management Journal* 6(4), pp. 314–324.
[9]B Andersen, J Howells, R Hull, I Miles and J Roberts (eds.) (2000) *Knowledge and Innovation in the New Service Economy*. Aldershot, UK: Edward Edgar.
[10]R Werner and U Wolfgang (2008) How to sell services more profitably. *Harvard Business Review*, May.
[11]http://productserviceinnovation.com/home/2010/04/18/xerox-services-to-survive-%E2%80%93-a-long-hard-road.

being driven by luck, Xerox grew from a deep cultural desire to satisfy its customers through its talented people who were determined to succeed.

The Copier Machine Company

In 1959, the company introduced the Xerox 914, the first plain paper photocopier which marked the first dramatic growth of Xerox in 1960s. The enormous growth was not only driven by product innovation but also by a thorough understanding of customers' needs. This led the company to pioneer the concept of "selling copies" to customers instead of "copier machines".

However, Xerox's dominant position in the copier market was challenged by fierce competition through the late 70s and early 80s. After experiencing failures in its attempts to commercialize new technologies (e.g., personal computing), the company had to rely heavily on its printing and copier after-market business as the main revenue source.

The Document Company

In late 1980s, Xerox decided to rebrand itself as "The Document Company", positioning itself as solutions provider through the entire document lifecycle. Besides copiers and printers, the company established its core advantages in colour digital printing technologies, document management systems, and services. A new services business unit was established, which specialized in outsourcing of mailrooms, production services, and fleet management. Unfortunately, these efforts were not profitable for Xerox. In spite of the rebranding efforts, it retained the tag of being a copier company.

A Services-led Partner

The above reality forced Xerox to outsource most of its manufacturing to Flextronics retaining only the high-volume equipment and specialist manufacturing operations. It implemented Lean-Six-Sigma to cut cost base and focused on value addition to customer processes. Execution of this ambitious transformation required a change of mindset across all aspects of the business — from Engineering & Production to Sales & Marketing and is still ongoing. The transformation included following aspects:

- A deliberate shift in hiring policies to recruit the right people with the right culture who understand customer value.
- An effective sales model which led to separating the services sales into a separate division, thus creating focus and a degree of internal competition for customers.

- The integration of services and products to drive profitable delivery of services.
- Commercial and operational management of services contracts which are often customized to the specific needs of the clients. (This remains a key challenge even within this sophisticated company.)

After many years of struggling in the transformation process, Xerox has become a "services-led partner in key enterprises" bridging the gap between paper and digital worlds *via* IT infrastructures. 70% of its revenue comes from post-sales services (i.e., reducing document costs for MNCs), which is now growing at 18% per year. 20% of revenues are contributed by a service annuity stream, which is not connected to the lifecycle of the product. Rather than the technology, it is this customer solutions and services approach that now differentiates Xerox from its competitors. Not only has this allowed Xerox to rebuild its margins, but has also provided the springboard to the new non-product-based lifecycle services that are yielding better revenue profits.

Today, the differentiating feature of any product is its service component, often regarded as the "make or break" factor. Services are more apt to be re-engineered than physical products and so a unique offering can be created, and recreated, in the market of commodities. The market for mobile phones, as a case in point, is unquestionably reliant on the service features associated with the device. Now more than ever, the operating system platform, to which a mobile phone company subscribes, is intimately tied to the potential success of the company's smartphone offering. Indeed, the service component of the value proposition of a business is playing an impactful role on the consumer's willingness to purchase. When mobile phones first came into the market they were so expensive that only businesses or the wealthy could afford the luxury of acquiring one. Paradoxically, one of the values of a mobile phone is the fact that it is a ubiquitous device which connects people irrespective of their location and time zone; but this would not have been possible if it had remained a niche market only for the higher-income customers. If the market had continued to view the mobile phone as a product, instead of considering the economic value of the telephone service that it provides, mobile phones would have never succeeded. Mobile phone operators today attract customers by out-subsidizing the price of the phone, but the real competition is in the pricing of the services linked to the service provider.

Experience Innovation

Servitization brings the concept of customer value to the forefront. Companies started to realize that it is the service component of a market offering that creates value addition and the comparative and competitive advantages. As John Kay surmises, "*the iconic manufactured goods of modern life are the iPod and the cruise missile, the Viagra tablet and the Coca-Cola can. For each, the resource content is a negligible part of the value of the product and so is the manufacturing cost. They are, in reality, packaged services*".[12]

Organizational Zoom ...

Coca-Cola: Availability of the "American dream"

Consider Coca-Cola, for example. During the Second World War, when the Coca-Cola Company strived to *"put a bottle of Coke in the hands of every American soldier"*,[13] the utility of the product had more to do with its availability in remote areas, and the experience of buying into the "American dream" than the physical product. The association with status also plays a role: Most people would rather be seen drinking a can of Coke than a supermarket branded version of the same product. The brand itself evokes an experience. Clearly, the utility derived from Coca-Cola goes far beyond the sensory aspect of the product, instead the value lies in the experience associated with consumption, the image, the brand, and the "American" experience.[14] The experience economy finally may create new ways to delight customers by taking a broader view of their lives and their interaction with the product, thereby delivering new, unexpected signature moments.[15]

Increasingly, companies are creating new values and gaining brand loyalty not by focusing on particular product features or design, but by

---

[12] J Kay (2010) Why you can have an economy of people who don't sweat. *The Financial Times*, October 19th.

[13] http://www.thecoca-colacompany.com/heritage/pdf/stories/Heritage_CokeStories_MILITARY.pdf.

[14] http://www.thecoca-colacompany.com/heritage/pdf/stories/Heritage_CokeStories_MILITARY.pdf.

[15] http://lippincott.com/files/documents/news/Lippincott_Perspective_Experience_Innovation_2014.pdf.

re-conceiving the broader experience of how customers use their products. Airbnb didn't redesign the travel portal or the hotels; it completely rethought how travelers can find the room they need. Car service company Uber didn't alter the vehicle or retrain the drivers; it fundamentally changed how people order, meet, and pay for the car service. Flipboard did not create new content, instead redesigned users' reading experience with its elegant format and personalized high-quality news and social content. And while Gillette developed one more blade on the razor, innovators such as the Dollar Shave Club arrived on the scene with a convenient, low-cost mail subscription-based model, fueled by social-media frenzy.[16]

Even the famous product innovation leaders realize that they can improve sales by enhancing customer experience. Nike Plus innovated the fitness experience for the community, not the shoes. Tesla's electric car is truly like none other, but Tesla does also offer to come to your home to service your car rather than making you go to a dealership. In each case, these experience innovators redesigned the customer experience, and not just the conventional product features. These companies solved customer problems in a way unique to their brands — with a rich array of experiences that surround and connect to the core offer. By innovating experience, companies can find untapped sources of differentiation to drive loyalty, preference, and growth.

Organizational Zoom...

NIKE Plus: The Personalization System

As a traditionally R&D focused company, NIKE is establishing a reputation for not just producing shoes, but also in building lifestyle. Over the recent years, the company showed a strong shift in its strategy from being a sportswear brand to enabling customized, personal experiences.

NIKE Plus system is a perfect example of this strategy. NIKE Plus Running is designed to map the running process of runners to sustain their motivation. The app tracks user's running distance, pace, and duration with GPS and calculates calories burned. Similar to a personal coach, it also provides audio feedback every mile as the user runs.

---

[16] http://lippincott.com/files/documents/news/Lippincott_Perspective_Experience_Innovation_2014.pdf.

It is easy to get demotivated when running. Peer recognition is therefore incorporated into NIKE Plus *via* Facebook and Path plug-in, which makes it easy for the user to share running routes and tag friends. The app also helps the user to push the performance upward. The tracking ensures a continuous feedback that allows the user to plan new forms and levels of exercise. There are badges to win for excellent performances.

Other innovative features such as weather forecast, "power songs" to motivate users to push their speed and endurance further, as well as an analytics web portal that allow users to read, analyze and share performance data on the PC, create an intelligent and innovative running experience for users.[17]

Trevor Edwards, VP Global Brand & Category Management, gives some numbers on the acceptance of the system: "35% of the members that we surveyed are actually new to using NIKE footwear. So, we've brought more consumers into our franchise . . . , 93% said they would recommend it to a friend, 93%. This is an incredible sticky proposition, a great way to build loyalty for our brand and obviously to build the business".[18]

Charlie Denson, NIKE's Brand President, describes the growth plans NIKE has with the system: "That is a dedicated consumer experience. It is changing the game, and it's creating that competitive advantage for us. We would like to see 15% of all runners using NIKE Plus, 15%. Now, now that's not a very big number, except for there's 100 million people who call themselves runners worldwide . . . "[19]

This confident statement stresses that this really has been a clever idea to provide customization in this industry in a rather simple way, but in one that matters to the consumer. And with a target of 15 million users, this would be one of the largest mass customization programs ever.

Emphasizing Customer Centricity

A key feature of servitization strategies is a strong customer centricity. Customers are not just provided the products, but also the broader and more tailored "solutions". These deliver desired outcomes for specific customers, or types of customers. This customer orientation consists of two elements. One is a shift of the service offering from product-oriented services to "users' processes-oriented services" (i.e., a shift from a focus

[17] http://www.innotour.com/innovationCases/2013/05/nature-as-a-fitness-center-the-nike-approach/.
[18] http://mass-customization.blogs.com/mass_customization_open_i/2007/02/the_consumer_de.html.
[19] http://mass-customization.blogs.com/mass_customization_open_i/2007/02/the_consumer_de.html.

on ensuring the proper functioning and/or customer's use of the product, to pursuing efficiency and effectiveness of end-users' processes related to the product). Another is a shift of the nature of customer interaction from transaction-based to relationship-based (i.e., a shift from selling products, to establishing and maintaining a relationship with the customer).[20]

Indeed, there has been a move by manufacturing firms toward "integrated solutions" as opposed to a single product concept. Windhal and Lakemond (2006) point out that an integrated solution, where the product is integrated with the service element, is more effective at addressing specific customer needs; this suggests a "*shift in focus from product functionality to the actual outcomes of their products and services for customers' operations and processes*".[21] Solution-based innovation focuses not just on the physical product functions and features built in hardware; but on accumulated company expertise, or "soft knowledge," as well. For example, a solution vendor might take responsibility for one of customer's subsystems or for an entire capability, such as IT. One of the fastest-growing groups within IBM Corp. is its global solution business, which often provides customers with front-end consulting to define the solution, later following up with systems integration using IBM software and hardware, and, in some cases, even providing the customer with financing. The capability to compete by offering solutions is based on some of the skills developed in the product space, but managers have to add skills in areas such as domain knowledge, supplier management, and pricing.[22]

The same trend — selling integrated solution rather than pure products — can also be seen in other industries. In healthcare, for example, many pharmaceutical companies are under significant pressure as the cost of developing drugs is increasing and many of the traditional drugs are coming off patent. So the generic manufacturers move into the market which has prompted pharmaceutical companies to rethink their business models — defining themselves as providers of healthcare solutions. Think like a patient — most of us don't want the products that pharmaceutical companies provide. We'd prefer not to be ill in the first place. So now, many health policymakers also argue that it's time to shift from a model

[20]R Oliva and R Kallenberg (2003) Managing the transition from products to services. *International Journal of Service Industry Management* 14(2), pp. 1–10.

[21]C Windhal and N Lakemond (2006) Developing integrated solutions: The importance of relationships within the network. *Industrial Marketing Management* 35(7), p. 806.

[22]CK Prahalad and V Ramaswamy (2003) The new frontier of experience innovation. *Sloan Management Review* 44(4), p. 12.

of treating ill health to one that promotes well-being. If someone can provide healthcare solutions, which reduce the likelihood of illness, the interests of providers and customers are again much more closely aligned.[23] Pharmaceutical firms can help facilitate such a turnaround by linking new services to their product, as well as by applying their business expertise to healthcare systems.

Organizational Zoom . . .

Johnson & Johnson; Pfizer: Seeking New Source of Growth by Selling Solutions for the Pharmaceutical Industry

During the G8 summit in London in December 2013, pharmaceutical and biotech industries participated in a discussion on how to meet the challenge of caring for an estimated 100 million people with Alzheimer's disease by 2050.[24] They participated not just as providers of drugs and devices that healthcare providers and patients might wish to buy, but also as partners in creating new healthcare systems.

It is a change that has been gradually taking place in the past decade. The global healthcare systems, whether they are in the UK or the US, India or China, commonly face increased demand in an environment where resources and particularly financial resources are constrained. In the UK, for example, the National Health Service (NHS) faces a projected spending gap of £30 billion a year by 2021, as the number of older people with multiple long-term conditions such as cancer, diabetes, and heart disease will increase inexorably. According to the World Health Organization Global Health Expenditure database, health spending in the US accounted for over 17% of annual GDP since 2010, while it reached only 4.6% for Singapore and 11.7% in France.

It is time to shift the focus from treating illnesses to promoting well-being. Pharmaceutical firms are crucial in bringing about such changes, which also brings new market opportunities where their knowledge and expertise would be the key factors to excel in healthcare service business.

Janssen is a good example. The company created the Care4Today master brand to reflect their commitment to transforming healthcare for the consumer.[25] This master brand offers a text service, through a two-way

---

[23] http://andyneely.blogspot.sg/2013/11/what-is-servitization.html.

[24] http://www.theguardian.com/sustainable-business/big-pharma-and-healthcare-providers-join-forces.

[25] Company's website. https://www.janssenhealthcareinnovation.com/care4today.

platform with an app and a website that can help patients and carers to take their medication correctly.

Another example, from the parent company Johnson & Johnson, is the text4baby service that provides mothers with health information *via* text message. The service started in the US and has been adapted by the Mobile Alliance for Maternal Action (MAMA) for use in the developing world, where mothers often have mobile phones but little or no access to healthcare professionals.

Pfizer, meanwhile, has launched a new service to help NHS organizations redesign their services to be more efficient. In 2011, for instance, the US pharmaceuticals company worked with the Velindre Cancer Centre in Wales to develop nurse-led follow-up clinics for patients with gynaecological cancer. Pfizer also provided the project management and data analysis to help develop a service for the NHS that eliminates long waiting times to see a doctor.

Undoubtedly, servitization is a transformation journey — it requires firms (often manufacturing firms) to develop the capabilities they need to provide services and solutions that supplement their traditional product offerings. To accomplish this transformation — to sell services and solutions — involves significant changes internally within many traditional manufacturers. They have to recognize that the product is a platform to deliver service. They have to build solutions that deliver the outcomes that will be desired and valued by their customers. In essence, these solutions are often captured in product–service systems, combinations of products and services. Customers only realize value from these when they actually receive the service — hence the concept of utility or value in use.[26]

HOW QUALITY OF LIFE DRIVES THE SHAPING OF EXPERIENCE

Another dimension to take into account in order to understand the rise of the services economy is the growing importance of the notion of quality of life as a central value in modern society.[27] Consumers, even organizations to a certain extent, pay attention to memorable experiences that ultimately

[26] http://andyneely.blogspot.sg/2013/11/what-is-servitization.html.

[27] Some materials for this development are taken from: H Mathe, X Pavie and J Egal (2012). *Quality of Daily Life: A Key Driver for Performance and Engagement in Organizations*. Paris: ESSEC-ISIS Cahiers Innovation & Society, March. ISBN 978-2-36456-049-9. Authors would like to express their gratitude to the many people who contributed to the discussion and to the organizations involved: Danone, British Ministry of Defense, University of Maryland, Asbury Communities, and Bangkok Hospital.

generate a feeling of fulfillment of a higher level of need in terms of quality of life. Before further discussing the processes of innovation in a service- and experience-driven modern economy, it is important to properly define the very notion of quality of life and to identify relevant attempts to actually measure it. Based on a wealth of business illustrations, we will then explain how to leverage natural aspiration for quality of life in formulating advanced products and service equations.

Rethinking the Concepts of Growth and Wealth

For most people, economic growth may not be perceived as generating direct improvement in the quality of their life. National income statistics indicators such as Gross Domestic Product (GDP) were built as measures of market economic activity, related to the notion of wealth of Nations. As their predominance on economic statistics became more vibrant, these indicators were increasingly assimilated to measure the societal quality of life, in a way that seemed to imply that economic development, through an increase in the GDP, would therefore create well-being for individuals. Today, this approach is seriously challenged. First, the bias results from the use of metrics based on medium incomes, i.e., the total wealth divided equally between individuals. The income inequalities between individuals keep widening in most modern societies and these inequalities are infringing the quality of life of people; generating for some a feeling of insecurity, unease, frustration, and anxiety. Other flaws in traditional economic models based on statistically measured growth can be identified, such as the difficulty to grasp the economic activities that take place within home and heavily contribute to individual well-being as much as, in fact probably more than, market consumption and production. Also, the standard GDP criterion does not take into account resource depletion and environmental degradation.

Awareness about the liabilities of traditional wealth indicators naturally leads researchers to question and challenge this state of knowledge. The normally available wealth indicators do not provide indication about the evolution of the well-being of the individual member of a society, neither about environmental risks. During 1970s, economists, such as Meadows,[28] Nordhaus and Tobin,[29] among others raised related issues. A new idea of wealth started to emerge from the debate; it included new dimensions

[28]DH Meadows (1972) *Limits to Growth.* Signet.
[29]WD Nordhaus and J Tobin (1972) *Is Growth Obsolete? Retrospect and Prospect. Economic Growth, Vol. 5.*

such as environmental legacy and social heritage. Then the concept of "sustainability" was finally introduced as a major guideline for the economy. The concept of Sustainable Growth was first defined by the Brundtland Commission of the United Nations in 1987.[30] Many new theories emerged from that work, including the "Triple Bottom Line" approach, which was amply discussed by Elkington later.[31]

Sustainable development, associated with social progress and well-being, now stands as an important notion in the economic vocabulary. The authors of the 2009 "Report of the commission on the measurement of economic performance and social progress", state that "Current well-being has to do with both economic resources, such as income, and with non-economic aspects of people's lives (...) Whether these levels of well-being can be sustained over time depends on whether stocks of capital that matter for our lives (natural, physical, human, and social) are passed on to future generations".[32] The Triple Bottom Line debate challenges the way companies address issues indirectly related to direct profit making process. The question of how the ecosystem, the customers, employees, or suppliers may actually be influenced by the organization is more vocally raised. Many observers defend that more people are now focusing on "being" than on "having". Words of progress, happiness, wellness, and personal development were added to the notions of social status and income per capita. The World Health Organization has defined quality of life as "individuals' perception of their position in life in the context of their culture and value systems in relation to their goals, expectations, standards, and concerns. Quality of life is a broad ranging concept affected in a complex way by the person's physical health, psychological state, level of independence, social relationships, personal beliefs, and their relation to the salient features of their environment".[33] Indeed, the quality of life is entirely focused on the individual as a unique human being, and seems to require people-centered policies.

---

[30] *Report of the World Commission on Environment and Development: Our Common Future* (1987). Oxford, UK: Oxford University Press.

[31] J Elkinson (1998) *Cannibals with Forks: The Triple Bottom Line of 21st Century Business*. Canada: New Society Publishers.

[32] JE Stiglitz, A Sen, J-P Fitoussi (2009). Report of the commission on the measurement of economic performance and social progress. In *French*: *Richesse des Nations et bien-être des Individus*. Paris: Odile Jacob.

[33] M Power and N Kuyken, on the behalf of the Whoqol Group (1998) The World Health Organization quality of life assessment (WHOQOL): Development and general psychometric properties. *Social Science & Medicine* 46(12), June 15th.

Organizational Zoom ...

Danone, Societal and Environmental Responsibility

Since 2008, Danone introduced societal and environmental targets for its 1,000 top executives' comprising general managers, members of management committees, and senior executives. On top of their fixed way, one-third of their total variable pay component is dedicated to meeting these targets.

The variable pay compensation is indeed calculated in three parts: One-third depends on meeting societal goals and achieving targets linked to the environmental and social impact of the business, as well as the ability to develop, promote, and train; one-third depends on achieving specific management targets; and the remaining one-third relies on global economic and financial targets.

Danone has also implemented societal and social entrepreneurship projects, which contribute highly to its positive image and culture, but which unfortunately may not involve many Danone employees on a daily basis. Among the well-applauded projects, the company teamed up in 2005 with Grameen Foundation of Bangladesh to develop yogurt for some of the poorest consumers in the world. The goal was to create a cheap product that would provide atleast 30% of the minimum daily requirement of key nutrients. The design of the Bogra factory had two aims: Keeping costs low and using non-polluting technologies.

Other social and societal projects at Danone include the Social innovation Labs and the Danone Supporting Life initiatives.

As a matter of fact, many commentators feel that obsession with economic growth has pulled away public attention from daily concerns about health, education, work, democracy, and environment. Factors influencing individual well-being differ and it would be irrelevant to impose any sort of universal norm. However, some aspects of well-being may be measurable, especially life expectancy, children mortality, nutrition, employment issues, crime, and corruption rate. Research to find the right criteria and measures to build a reliable indicator has been conducted all over the world, leading to various indexes used at national levels.

Attempts to Measure Quality of Life

The David Morris' Physical Quality of Life Index (PQLI) is among the first criteria introduced during the 1970s and is based on basic literacy, infant mortality, and life expectancy. The concept of Gross National Happiness was coined in 1972 by Bhutan's former King Jigme Singye Wangchuck. It is based on four pillars: promotion of sustainable development; preservation and promotion of cultural values; conservation of the natural environment; and establishment of good governance. The Human Development Index (HDI) emerged in 1990 as a composite statistical indicator of human development combining measures of life expectancy, education, and standard of living. Designed by economists Mahbub ul Haq and Amartya Sen, this was an attempt to quantify the options available to individuals in a given society. The Genuine Progress Indicator (GPI), or Index of economic Well-being, by Lars Osberg and Andrew Sharpe, aims at measuring whether a country's growth, increased production, and expanding services actually result in the improvement of welfare and well-being of the people in the country. Eventually, the Happy Planet Index was introduced by the New Economic Foundation in 2006 to challenge the GDP and HDI indicators as the Foundation felt that the other indicators did not take sufficient account of the sustainability imperative quotient. Since 2008, the Gallup-Healthways Well-being index evaluates emotional and physical health, present and anticipated life situation, work satisfaction, and access to basic necessities, through daily phone survey of at least 500 individuals in the US. Over two million interviews have been conducted between 2008 and 2015. The daily results of 100-point measurement system allow tracking down the factors that impact individual well-being.

Organizational Zoom ...

Bangkok Hospital — BDMS group

Bangkok Hospital is a private institution established in 1972 and handles one million patients a year. The flagship facility is the largest private hospital in South-East Asia and is regarded the most technologically advanced hospital in Thailand. It treats over 20,000 foreign patients a year. The hospital chain boasts of patients from several countries such as Japan, UK, UAE, USA, and Germany. To better serve this very diverse population, the hospital employs interpreters in 22 languages in addition to English-speaking staff, and provides connectivity to enable their international patients chat with the doctors on-line.

Food service is also among the top priorities for the hospital. Indeed, patients expect to be served a quality meal that meets the nutritional requirements of their medical treatment. Food has to be provided at the right time and at the right temperature in the rooms. The choice of menu depends upon the diverse cultural origins of the patients. Some overseas patients may be willing to experiment Thai cuisine during their stay, whereas others may prefer to stick to the traditional tastes they enjoy in their homeland. Nutrition as a key lever for quality of daily life is part of the hospital's culture, even for the staff, nurses and physicians, who are encouraged to keep an eye on calories.

A lot of initiatives are taken in Bangkok Hospital to improve the patient experience. For instance, in order to reduce perceived waiting time, classical music concerts are organized in the main lobby.

Quality of life, well-being and societal development are currently appraised with indices referring to material living standards (income, consumption, and wealth), health, education, personal activities, political voice and governance, social connections and relationships, present and future conditions of the environment, insecurity as well as physical nature. Yet, it is also recognized that measuring well-being involves, at least for a part, subjective criteria, i.e., asking people how they feel about their lives. In 2009, the Stiglitz commission concluded that 'research has shown that it is possible to collect meaningful and reliable data on subjective well-being (...) and a better understanding of its determinants'.[34] The report suggests some directions to possibly appraise individual well-being instead of the general output summarized by GDP. In early 2000, a model was suggested to grasp some sense of daily quality of life as perceived by individuals. The Diener's model introduces cognitive features (work, family, and financial condition); and presence and absence of positive emotions.[35] The Center for Well-being at the New Economic Foundation suggested a model in 2011 based on three dimensions:

— People's satisfaction with their life, generally measured by an indicator called life satisfaction that captures satisfaction, pleasure, and enjoyment.
— People's personal development for which there is no standard psychological indicator yet. The concept includes being engaged in life, curiosity, flow (a state of absorption where hours pass like minutes), personal development and growth, autonomy, fulfilling potential, having a purpose in life, and the feeling that life has meaning.
— People's social well-being: A sense of belonging to our communities, a positive attitude toward others, a feeling that we are contributing to

[34] JE Stiglitz , A Sen and J-P Fitoussi (2009). *Op. Cit.*

[35] E Diener (2000) The science of happiness and a proposal for a national index. *American Psychologist* 55, January.

society and engaging in pro-social behavior, and believing that society is capable of developing positively.[36]

Even if the reviewed models start focusing on people and tend to integrate subjective measures, the insights they provide only focus on quality of life from a macro-economic perspective. They do not unveil the drivers of quality of life in collective environment where they matter every day. At that stage, they fail at suggesting guidance for organizations to influence the quality of life of people in their ecosystem.

Leveraging Quality of Life and Formulating Creative Services

Because of the importance of interpersonal interactions, collective environments play a major role in individuals' quality of life. Therefore, it makes sense to suggest that organizations should address the issue of the quality of life at work and other forms of group life. Whether it proves feasible to create a comprehensive index for quality of life in organizations, or not, taking into account various subjective factors that can influence its perception opens up opportunities. Understanding the drivers of quality of life can help organizations to improve people's experience and satisfaction, favor individual progress, and therefore enhance individual engagement as well as overall performance. Optimizing performance, while encouraging people engagement, is the main challenge for any organization. Numerous studies have shown that healthy individuals tend to be more productive than unhealthy ones, and obviously physical condition is far from being the only driver of performance. Yet, contemporary developed economies seem to be plagued by high unemployment and wage inequalities, pressure, stress, and often depression. Then does not addressing the issue of quality of life (kind of) sound like an ironic utopia?

Organizational Zoom . . .

Colchester Garrison — The British Ministry of Defense

With 8,000 military and civilian personnel, Colchester Garrison is the largest brigade in the British Army. For the military men, physical

[36]The Center for Well-being, NEF (2001). *Measuring our Progress: The Power of Well-being.*

> environment can be seen as secondary, after health and safety that often come first among the priorities. There is a strong vocational incentive when people enroll in the army and the experience they are looking for does not include expectations for great comfort, gardens, and architecture. Yet when they get back from the field, those elements prove more valuable to them, as well as the need for leisure and socialization. However, what may appear as an improvement in terms of quality of life, such as a single room for each soldier, might not prove beneficial to them considering the psychological trauma they might come back with, after six months of fighting on the field.
>
> Also, the situation of the direct families provides an illustration regarding the "emotional" dimension. It is most significant when military personnel is sent in operations, for up to six months to Afghanistan, for instance. The direct families usually remain in the garrison, feeling anxious about spouses or parents fighting overseas along with high levels of stress. At that stage, dependents demonstrate strong relational need, as they often want to share their emotions and relieve the pressure. But when the spouses are back, mostly men, the 'vocational' dimension comes to the fore, since the wives feel that their personal identity is only linked to the status of their husband. They express the need to be recognized as individuals and not only as 'dependents'. Consequently, they look for fulfilling jobs, to develop their skills and networks, and to improve their self-confidence and self-image.

Providing insurance, training, and improving ergonomics is no longer sufficient. Quality of life implies a broader approach. It becomes necessary to pay attention to the practical difficulties that people face in their everyday lives that can vary from commuting to taking care of the kids, and more. The broader quality of life implies rethinking the organization, so that it allows people to fulfill their potential, and provide purpose, and meaning to their everyday life. We have suggested a methodology to define, analyze, and improve quality of life in organizational environments, considering different segments of activity and different categories of players within organizational ecosystems. The necessity to address the concept in aggregate implies consideration for personal and contextual dimensions. Individual perception of the quality of daily life can be influenced by numerous intimate characteristics: Culture, religion, social background, education, and personality. The influence of individual factors, subjectivity, and personal preferences on the assessment of quality of life proves complex. The roots of well-being and positive psychology date back to the 1940s, when Abraham Maslow introduced the notion of hierarchy of needs and of optimal life through the fulfillment of innate needs. The personal dimensions of an individual could then be summarized as a set of needs and expectations that influence his/her perception of the environment in which he/she lives. Five categories of personal dimensions may be suggested thus:

- Physical: Healthy body, usage of physical capacities.
- Relational: Intensity and quality of social interactions.
- Emotional: Psychological condition, feelings and sensitive drivers for action.
- Intellectual: Mental activity, stimulation, and engagement of the intellect.
- Vocational: Meaning of today efforts, improvement of capacities for action.

Personal dimensions are impacted by several contextual factors that can be economic, political, natural or environmental. Organizations are not always in a position to manage the negative impacts of exogenous factors, or be considered responsible for the constraints due to external factors. But common factors can be identified in an attempt to conceptualize the quality of life in an organization. Such common factors can be defined as five distinct pillars that might not have the same importance for every individual. However, collective environments usually favor similar behaviors in their ecosystem, for the common needs of their inhabitants. Somehow, any individual expects to reach a certain standard for each of these pillars in order to appreciate the quality of his/her daily life:

- Space: Room in the organization, and in the direct environment.
- Health: Conditions favoring the strengthening of the body and its functions.
- Social life: Accessible networking and socializing opportunities.
- Easiness: Feeling of fun, relaxation, and relieved pressure.
- Aspiration: Personal achievements and future prospects.

Achieving the minimum standard required by an individual for each pillar is the only key to avoid dissatisfaction in the overall quality of life. Dissatisfaction with one or more important pillars can drastically decrease the overall perception. Over-satisfaction with one or more pillars does not necessarily compensate for the weakness of one or more important ones. The question of dissatisfaction or satisfaction may be compared to the Motivation–Hygiene theory that Frederick Herzberg in the 1950s applied to the working environment as a concept of motivation and job

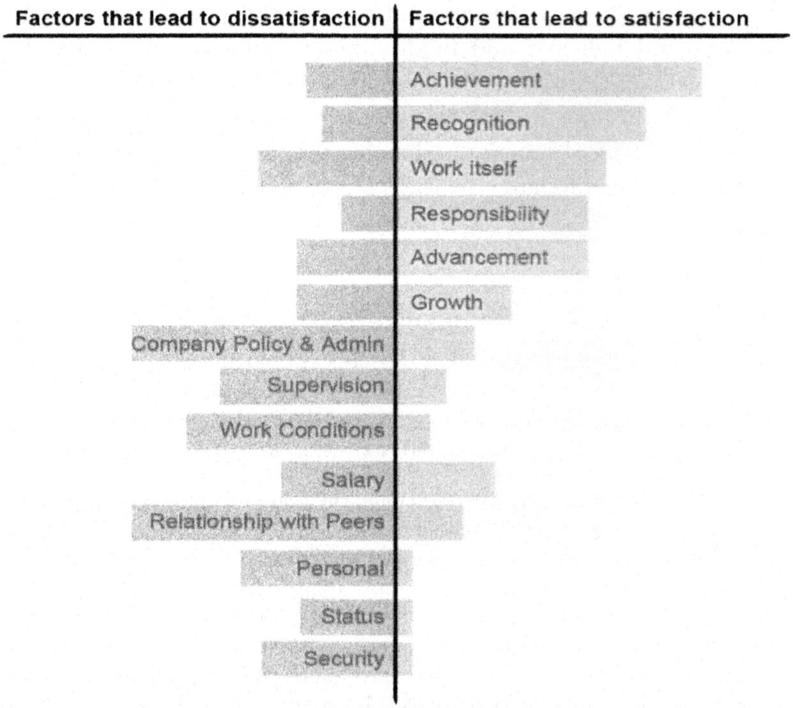

Figure 2.1: The 14 Factors Influencing Job Satisfaction/Dissatisfaction. *Source*: F Herzberg (1968) One more time: How do you motivate employees? *Harvard Business Review*, January–February.

satisfaction.[37] The Two-Factor theory states that hygiene needs, such as salary or working conditions, are not strong motivational factors but rather are more significant for preventing job dissatisfaction. Motivational factors stimulate satisfaction when they are fulfilled. Recognition, for instance, stands as a strong driver for job satisfaction, but if inexistent, it does not necessarily cause job dissatisfaction. Later, Herzberg identified 14 factors influencing job satisfaction/dissatisfaction as indicated in Figure 2.1.

In order to improve the perception of quality of life of individuals, employees or beneficiaries of its services, organizations can act on a series of levers, such as: Design of the working environment; learning opportunities, counseling, and coaching; health and safety standards, maintenance, and cleaning; technology advanced equipment; food service; sports and wellness

[37]F Herzberg (1959) *The Motivation to Work*.

facilities and supervision; events and get-togethers; commuting facilities; family care and benefits; and finally, opportunities to contribute to the society. It is therefore important for organizations to understand which lever they should activate, relative to any critical pillar of quality of life on which they want to generate a direct influence and improve individual's experience and satisfaction.

Organizational Zoom . . .

Smith School of Business — University of Maryland

Founded in 1921, Smith School of Business at the University of Maryland offers wide array of qualifications from bachelor degree to MBA and Executive programs. A top flying institution, it hosts 4,300 full-time students, 150 faculty members and a large staff.

Prior to the recent real estate additions, the issues of space, construction, and design were particularly challenging for the Smith School since there were different reasons for dissatisfaction among students and faculty members. Students felt there were not rooms available for group work. Some felt that the physical environment was a hindrance to the collective achievements of the group, and hence to their own achievement and future prospects. Faculty members were more concerned with the social aspect of the quality of their daily life, complaining about the lack of a proper lounge to facilitate networking and social interactions. The consequence was that the sense of belonging seemed to be missing.

For students, faculty, and staff, good sports facilities were accessible within the University of Maryland campus, and yet, they didn't seem to participate much in the activities. Lecturers reported they did not want to be closely involved in sport activities with students; while students felt they did not have enough time for their journey from the school to the sports facilities. Ideas were suggested to improve the situation: A private and specific sports schedule for faculty and a regular bus shuttle within the campus for students were established.

Chapter 3

THREE KEYS TO MAKING SERVICE INNOVATION A SUCCESS

The expansion of services certainly stands as the most significant and spectacular business trend in the past three decades. Nowadays, service delivery represents approximately 80% of the US Gross Domestic Product (GDP) and over 75% of the GDPs of most of the Organisation for Economic Co-operation and Development (OECD) countries. Increasingly, there is evidence that services dominate global economies and drive growth. Services are often referred to as the reciprocal of manufacturing because of the importance of labor in the total cost of most of service outputs, as opposed to that of materials in tangible goods. The expansion of services is the most spectacular business trend in the past three decades. Western economies, in particular, continuously add positions in services but shed jobs in manufacturing. After a long period of stagnation due to the predominance of manufacturing, international trade in services is now dramatically taking off. Wealth is finally created in the service economy, thanks to the "capitalization of innovation". Leading brands have emerged in services; and many are strong enough to carry an astonishing value. As it is frequently emphasized, "modern economies are inescapably service economies".[1]

[1]F Gallouj and F Djellal (2010) *The Handbook of Innovation and Services: A Multi-Disciplinary Perspective*. Cheltenham: Edward Elgar.

In a white paper based on the 2007 Cambridge Service Science, Management and Engineering Symposium, researchers defined services as "an economic activity offered by one party to another, most commonly employing time-based performances to bring about desired transformation results in recipients themselves or in objects or other assets for which purchasers are responsible" (...) "In exchange for their money, time and effort, service customers expect to obtain value from access to goods, labor, professional skills, facilities, networks and systems; but they do not normally take ownership of any of the physical elements involved".[2] In general terms, service operations are defined as those which provide value through information, savings in time, desired psychological state or experiences, or changes in the physical attributes or property. Services are produced by producers acting for recipients; or recipients providing part of the labor; and/or recipients and the producers creating service in interaction. One of the definitions of service systems is suggested as "dynamic configuration of people, technologies, organizations, and shared information that create and deliver value to customers, providers and other stakeholders".[3]

Indeed, services touch our lives every day. Today, a massive proportion of consumers show much more interest in the way a purchased good or service will satisfy their needs, than in the physical traits or technical performance of the goods and services themselves.[4] Customer service, on its own terms, is naturally credited with growing importance in a society dominated by the consumption of intangible value. Yet, it is increasingly difficult to satisfy the needs of specific and often quite different customers. For instance, in the medical equipment industry, there are very diverse types of "customers", such as a modern hospital in Zurich, an emergency health center in a developing country, an individual professional in Boston as well as the British army medical corps. Different customers require not only a range of dissimilar tangible products, but also the definition and delivery of a variety of different services — often based on idiosyncratic relationships with these customers. Today, globally most manufacturing and/or service

[2]University of Cambridge Institute for Manufacturing (IfM) and International Business Machine (IBM) Corporation (2008) *Succeeding through Service Innovation: A Service Perspective for Education, Research, Business and Government.*
[3]*ibid.*
[4]Dong-Sung Cho (2008) Four phases of the design revolution: Expanding the design domain and developing design theory. In *Designing Information and Organization with a Positive Lens: Advances in Appreciative Inquiry, Vol 2.* Elsevier.

companies are focusing on customization of responses to satisfy different requirements of different customers.

SERVICE INNOVATION PROVES CENTRAL TO IMPROVING ECONOMIC PERFORMANCE

Innovation in services, or through services in manufacturing industries, is of paramount importance. At stake is the ability of the companies to align their entire value producing mechanism with rapidly changing, consumer centric and information intensive, challenging, business environment. Innovative service strategies are now established on the basis of structured approaches. The adoption of technological advancement and the stimulation of competition have dramatically improved productivity in agriculture and manufacturing sectors over the years, significantly reducing the proportion of the working population contributing to these sectors. Services, on the other hand, are often perceived as low productivity, high labor-intensive and stagnant-in-performance activities. The criticism is not entirely unfounded over a 50-year period of observation. However, the profile of the 21st century high tech service firms, as well as the landscape of the entire service industry, is rapidly and significantly positively altered by the massive deployment of innovative formulae.

Innovations in services may vary from country to country, based on sector and organization size, however, the emergence of novel solutions in service product design, process engineering, shaping of physical and virtual environments, and reinvention of business models can be seen everywhere. In its usual accession, service innovation relates to new service solutions, introduced at any level right from the concept stage until the delivery of the end product, that positively alter the recipient–provider interactions in enhancing the beneficiary experience while creating economic value. Service innovation may be introduced by service firms but also by manufacturing companies that are aiming to include the provision of intangible value to customer, or government agencies addressing critical societal issues in relation with mobility, education, health care or security, among other topics.

Enhancing Beneficiary Experience and Creating Value to Society

Opinions on the ability of service organizations to innovate have been expressed in multiple circles. "Services are more innovative than previously thought", said OECD experts in 2005,[5] "In some areas, they are more innovative than the average manufacturing industry". In 2008, a group of researchers led by Mary J Bitner observed that: "In practice, innovation in services is less disciplined and less creative than in the manufacturing and technology sectors".[6] The understanding of how service firms actually innovate is also often questioned since the usual approaches developed in order to cope with the reality of the manufacturing industries don't quite apply in service sectors.[7] Researchers from Denmark have observed that "Until recently, services innovation has been overlooked in terms of policy development. The links between public policy and services innovation have largely been unexplored and underexploited".[8] According to Henry Chesbrough, "Service Innovation may disrupt traditional channels to market (...) to enhance significantly customer experience in a way which impacts upon the value chain as a whole. In this way, it is shaping emerging sectors, industries, and markets and contributes to structural change and industrial modernization".[9]

Investments in Research and Development (R&D), the traditional engine of innovation in manufacturing, might prove less important in service innovation than having highly skilled workers and strong relationships with other businesses. That said, the "Digital Disruption" associated with Information Transparency, New Media, Platform Makeover, and Smart Virtual Assistants unveil formidable opportunities for technology-driven service innovation.[10] Finally, factors such as harnessing the power of social media, putting a premium on relationships and tapping the viral capability of social networks as distribution channels, revolutionize the service sectors. Burberry's successful digital transformation, for instance, opens paths for breakthrough approaches to innovations in marketing in the luxury

[5]OECD Directorate for Science, Technology and Industry (Oct 2005). *Promoting Innovation in Services.* JT00191834.

[6]MJ Bitner, AL Ostrom and FN Morgan (2008) Service blueprinting: A practical technique for service innovation. *California Management Review*, 50(3).

[7]National Endowment for Science, Technology and the Arts (2008) *Taking Services seriously: How policy can stimulate the "hidden innovation" in the UK's service economy.* NESTA Research Report.

[8]Danish Institute of Technology (2012) *Service Innovation as a Driver of Growth.* Report under the supervision of N Sondergaard Laugesen.

[9]H Chesbrough (2011) *Open Service Innovation: Rethinking your Business to Grow and Compete in a New Era.* San Francisco, CA: Jossey-Bass-Wiley.

[10]*Technology Innovations Powering 21st Century Business* (2008) Leading Edge Forum under the supervision of P Gustafson and W Koff. CSC, Falls Church.

segment, whether it is about launching new fragrances on Facebook, or developing the "Tweetwalk" with Twitter, or being amongst the first brands on Google+.[11]

Numerous examples have demonstrated that innovation may actually reinvent the service firm based on the delivery of totally new customer experiences; it actually may lead to the reinvention of the entire service sector as well. In less than 10 years, Expedia has taken over the leadership position in the travel business that was earlier held by American Express for almost 150 years; previously ClubMed had introduced a radically new form of vacation concept that revolutionized the tourism industry. Southwest Airlines, IKEA, or McDonalds, amongst many others, certainly rewrote the rules of the game in air travel, furniture design and retail, and restaurant business respectively.[12] Even higher education systems and universities globally have to drastically re-engineer themselves in order to cope with the challenges of the modern world, as stated by Clayton M Christensen and Henry J Eyring.[13]

Innovative service strategies are now established on the basis of structured approaches. The role of customer service must be articulated during the formulation of the business strategy, especially in manufacturing settings where the building of competitive advantage is expected. Defining the mix of services to be provided requires the understanding of customer needs throughout the entire lifecycle of the customer/supplier relationship. To be successful in providing those services, management must recognize that service innovation is a corporate-wide activity, mobilizing enabling technologies and requiring careful cross-functional coordination.[14] Every functional division has a critical role to play. Inadequate balance between service function "integration" relating to coordination, and "separation" relating to specialization, creates conflicts among the different parties involved in the "service value chain", i.e., marketing, sales, R&D, production, distribution, and after-sales support. Providing conceptual frameworks

---

[11]Burberry's digital transformation, an Angela Ahrendts interview (2012) In *Digital Transformation Review,* 2. Paris: Capgemini Consulting.

[12]H Mathe (2012) Leading in Service Innovation: Three Perspectives on Customer Service Value Delivery. In P Wieser, F-L Perret and C Jaffeux (Eds.) *Essentials of Logistics & Management.* Boca Raton and Lausanne: EPFL and CRC Press, Taylor & Francis Group.

[13]CM Christensen and HJ Eyring (2011) *The Innovative University: Changing the DNA of Higher Education from the Inside Out.* San Francisco, CA: Jossey-Bass-Wiley.

[14]H Mathe and R Shapiro (1993) *Integrating Service Strategy in the Manufacturing Company.* London: Chapman & Hall.

and sets of implementation guidelines to manage the different dimensions of the "value proposition" were fashionable topics in business literature during the mid-90s. Currently, service innovation is definitely reinforcing its pre-eminence as a key activity for manufacturing companies operating in economically developed countries, for traditional or new service firms flourishing everywhere, as well as for government agencies all over the world.

> 'As Antonio Tajani and Johannes Hahn, both from the European Commission, claim in the introduction of "The Smart Guide to Service Innovation": "To achieve sustainable growth and jobs across Europe, a broad concept of innovation will need to be followed (...). Service innovation has an important role to play in such strategies. This should in particular contribute to providing small and medium-sized enterprises with favorable regional eco-systems and helping them to be competitive in global value chains. Service innovation should be seen in a strategic and integrated manner in order to shape new industries and restructure old ones".[15]

Understanding the Drivers of Service Innovation

Motivation to innovate in a service-driven business environment is frequently translated into willingness to Enhance responsiveness in organizations and formulate simpler solutions; provide personalized operational answers to identified needs; and finally, to formulate new and efficient economical equations. A few examples of inspiring experimentations are illustrated in the following sections.

Zara became the market leader by strategizing and implementing its "hyper fast fashion" supply chain concept, which enabled it to achieve unequaled performance in managing complex service systems in enhancing its operations responsiveness. In 2006, in the video game industry, Nintendo penetrated a large untapped market by launching the Wii family game that showcased the success of a simple experience-delivery based device. In 2007, Johnson & Johnson, which produces the leading anti-anemia "Procrit" medicine, unveiled a new website www.chemicare.com, dedicated

[15] *The Smart Guide to Service Innovation* (2013) A Tajani, Vice-President of the European Commission, and J Hahn, responsible for regional policy at the European Commission (Eds.). Brussels: European Commission.

to coaching caregivers in supporting patients in post-cancer therapies. Unexpectedly, this website became a huge success and is particularly an affordable source of key data on collateral effects, where the word of mouth has impacted in accelerating the development process for new treatments. In 2008, Novartis, also a leader in the pharmaceutical industry, launched its own Internet coaching solution.

Today, affordable and simple-to-understand well-being products have been introduced into the markets that effectively address the intrusiveness displayed by earlier products. For example, in 2010, Withings introduced a number of a weighing devices that could check for weight, BMI, body composition, air quality and so on. Intel and GE unveiled their Magic Carpet prototype that analyzes the health condition of the sick and elderly people based on their walk on a wired carpet system. If there is a deviation from the regular routine, the sensors in the carpet notice this and send out an alert notice via iPhone to their caregivers. In another industry, Progressive, a leading car and motorcycle insurer in the US had experienced dramatic growth during the 90s, when it introduced its "immediate response" and "comparison quote" processes. By dispatching an expert equipped in damage appraisal and immediate clearing of insurance appeals, to maximum number of accident sites, Progressive provided a simple and extremely responsive mechanism to customers when they needed such services most.

There are strong trends toward the necessity to address individually expressed requests, in consumer as well as in business markets, which are fueling infinite series of experimentations for innovative service delivery. To a certain extent, automation and self-service technologies, initially mobilized for productivity improvements sometimes at the detriment of customer satisfaction, now contribute to the shaping economically viable and emotionally valuable services to consumers. Tour operators, such as Atalante as early as in 1996 or Terra lncognita and Continents Insolites in recent times have focused on shaping completely personalized trekking and unique vacation experiences.

The spectacular success of the Subway franchised network, which became the largest restaurant chain in 2011 in the Quick Service Restaurant industry with more locations than its competitors, is based on its capacity to stimulate and accommodate individualized requests. In 2006, Swarovski started to post on its website a tool box inviting its customers to individually design their own piece of jewelry, while Timbuk2 was launched in New York City to deliver customized travel and handbags within two to three days timeframe. In the business to business era, Li & Feng from Hong Kong has become a leading logistics service provider by offering customized supply chain support to manufacturing, trading or retail firms. Also, Taiwan Semi Conductor Manufacturing Corporation (TSMC), has introduced the concept of the "foundry as an open wafer fab" to cope with individualized requirements from external product developers. In 2010, TSMC unveiled its "open innovation platform" to further engage each customer in the process of complex design.

The search for new business equation also proves a powerful driver in service innovation. When Blockbuster transformed its model of video delivery to renting shops from a flat \$60 per video charge to a more appealing \$5–\$10 with additional fees up to 50% of sales, its market share in this declining business grew from 25–38%. Since 2010, in the aircraft engine business, Rolls Royce gleans more than 50% of its revenue from servicing its engine fleet,[16] and is also credited to have re-invented the "power by the hour" program. Airlines remunerate the manufacturer, still retaining engines' ownership, in function of the number of hours of the engines real work. Based on the same logic, Caterpillar charges equipment users by volume of earth moved and Bombardier has started experimenting with a system of charge by mile covered by railroad equipment.

In order to stem the decline of its traditional copying business, Xerox has developed the 'managed print service' system. The dramatic shift from manufacturing to services in traditional industries is well exemplified in the iconic example of IBM which increased its services revenues from US \$10 billion in 1990 to 60 billion by 2010, while sales in hardware decreased from 45 billion to less than 18 billion in the same period. However, the company massively invests to sustain its technological leadership,[17] and to stay in line with its long lasting strategy of close customer and supplier relationships.[18] Over time, these relationships have motivated IBM to design and deliver advanced services, leading to the entire reshaping of its value equation.

---

[16] *Manufacturing, a tale of two industries* (2011) The Economist, July 30th.

[17] R&D spending in 2013; US\$ 6.3 billion for a total revenue of US\$ 104.5 billion (2013) *The 2013 Global Innovation 1000 study: Navigating the Digital Future* under the supervision of B Jarulselski, J Loehr and R Holman. New York: Booz & Company.

[18] *Briefing IBM, 1100100 and counting* (2011) The Economist, June 11th.

Leveraging Engines for Successful Services Development

In order to reach their objectives of providing more responsive and more customized services while adopting new business models, companies tend to invest in customer-centric policies with newer design thinking activities, and advanced technology. Naturally, web-based services perfectly fit in the paradigm of service innovation, but creative answers emerge from all sides of the service sector whether or not they are related the digital age.

Organizational Zoom . . .

Emirates, Airbnb, Decathlon, Hyundai, and Olive Garden

Emirates Airline provides one of the perfect examples of following a customer-centric approach. It offers 1,000 entertainment-on-demand channels sourced from over 15 countries including movies in 20 languages. It has been repeatedly elected as the World's Best Airline. By generating the renting of more than 50,000 accommodations a day, Airbnb also provides a truly customer-centered service although in a very different way. By acquiring the participation of committed sports people in the design and retail of sports clothing and equipment, Decathlon does give the customer a lot of attention at every stage of the value chain. Similar idea went with Hyundai which experienced growth in the US and European car markets in 2008–2009 despite the deep recession in these economies. Hyundai achieved this by extending innovative insurance policies including the very popular 'income loss' coverage. Olive Garden, part of the Darden group, became the largest chain of Italian-themed full-service restaurant in North America through the deployment of their "hospitaliano" concept that aimed at reaching for 100% guest delight. They also developed a unique Italian dining experience that starts right from their interactive website, where recipes can be downloaded along with free culinary training. In addition, a lot of attention was paid to the design of the physical facilities, shaped as Tuscany country manors and equipped with genuine furnishing.

On the other hand, design thinking is pursued in order to reassess the way new services convey messages. Roberto Verganti, formerly at Harvard and now back in Italy as a Professor with Politecnico di Milano, defends that design-driven innovations do not come from the market: They create new markets. Verganti believes that companies have to go beyond the usual observation of users and customers to instead look for "interpreters" who he defines as "The community of players, from artists to technology suppliers to design schools, that surround every product (or service) and deeply understand and influence how people give meaning to things".[19] The iconic example of Alessi's — manufacturer of household items — "family follows fiction" project in 1993, illustrates the appealing power of radical innovation of meanings in physical objects. "This unsolicited meaning turned out to be exactly what people were looking for" says Verganti, "... meanwhile, Alessi has enjoyed double-digit annual growth".[20]

Many situations may be identified in the service sector as well, from the introduction — long time ago — of the concept of central atrium architecture by Hyatt hotels chain, to the attention given by Nespresso to entirely reshape the coffee consumption experience, ..., and meaning. Umpqua bank, headquartered in Portland, Colorado, drastically readdressed the retail banking experience by transforming its branches as almost community centers dedicated to the development of the neighborhood they belonged to, while transferring all their operations to the Internet. Part upscale hotel, part retail, Umpqua's innovative new store, designed by Ziba, invites customers to read the paper, enjoy a free cup of coffee, surf the internet, and potentially shop for banking products. The new allocation of space has helped create cultural cohesion around the Umpqua vision. Naturally, attention to design is instrumental in industries such as theme parks, entertainment or video gaming, where a well-developed product such as Resident Evil can sell 45 million units.

Powered with technology, the "Pass Navigo" system in the Ile de France region has significantly improved public mobility by allowing one universal electronic pass to be accepted as a form of payment by all operators of transportation means in the area, from subway lines to bicycle renting. However, implementation of Near Field Com — the smartphone universal payment system — has not proved as successful as expected, but it is expected to take off in near future. In Soho London, the inamo restaurant introduced 'E-Table', an interactive ordering table for food and drinks that also serves as a virtual table cloth, taxi or entertainment booking booth and a connection to the shopping possibilities in the vicinity.

---

[19]R Verganti (2009) *Design-Driven Innovation: Changing the Rules of Competition by Radically Innovating what Things Mean.* Boston, MA: Harvard Business Press.
[20]*ibid.*

Customers from their tables can even watch the inamo chefs working in the kitchen. Automatically monitored waiting time is limited to 15 minutes. This formula leads to a higher customer turnover rate, higher revenue due to greater incentive to spend, lower staff cost and seamless integration of the restaurant management system with the point of sale system. From education — with the recent explosion of the massive open online courses — to basic translation services available from any smartphone, to stock brokering, technology is altering all sectors of the service industry and dramatic examples of innovative technology-powered solutions can be found all over the world. However, while technology definitely facilitates the implementation of breakthrough delivery solutions in services, it is not a substitute for innovation strategy.

RETHINKING PROCESSES TO CO-CREATE SMARTER SOLUTIONS WITH CONSUMERS

There are several commonalities in the way successful companies in different service industries address innovation. Firstly, many successful innovative service companies address the market considering three different forms of interrelated, yet distinct, targets: Ultimate beneficiaries of the package whose willingness to enter the relationship is the key; paying bodies who inject most of the financial resources in the system, and entities or individuals who "prescribe" the consumption of the given services. For example, users in the cities will directly benefit from JCDecaux series of services, from bus shelters to public toilets and bicycle renting, while advertisers will entirely cover the cost, and city authorities will "prescribe" beneficiaries' access to the value when awarding exclusive license to JCDecaux in a given geographical area. CS2 Lawyers, an early adopter of automation and technology adoption in professional services, works efficiently to provide timely and information rich legal support to needy motorists, however, the expenses are taken care of by insurance companies, who save on administrative costs as well as enhance customer satisfaction. Maximiles provides benefits to end consumers who collect miles from multiple sources and convert their earnings into gifts or free services. For managing this multi-brand loyalty program, the company gets paid by the brand managers. Triselec, in the North of France, specializes in selective trash conversion while offering jobs to non-qualified people to help them return

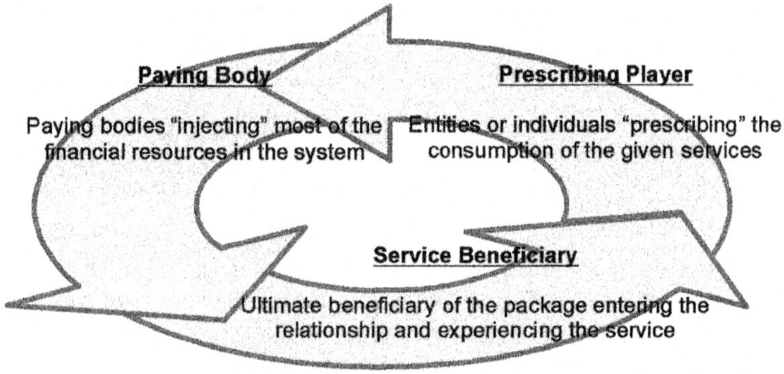

Figure 3.1: Three Levels of Receivers of the Service Value.

to the job market. However, Triselec could not have operated successfully without the agreement of local and regional authorities. Financial resources are obtained, thanks to the sales of selected items of recyclable value, while the ultimate beneficiary involves himself into sorting of trash at home and office in order to participate in what is perceived as an act of responsible citizenship.

Engaging Three Categories of Value Recipients

Many successful innovative companies address the market considering three different forms of interrelated, yet distinct, targets for their service offerings (Figure 3.1):

- Service beneficiary: Ultimate beneficiaries of the package whose willingness to enter the relationship is the key.
- Paying body: The paying parties who inject most of the financial resources in the system.
- Prescribing player: Entities or individuals who somehow "prescribe" the consumption of the given services.

Organizational Zoom . . .

JCDecaux: Addressing Service Recipients as Citizens and Consumers

In 1964, a new company was launched in France by Jean-Claude Decaux with the innovative concept of providing bus shelters in cities, free of charge, funded entirely by first-class advertisements. Next, it introduced display of city plans on the advertising panels, and by 1976, the first time-share of multi-brands panels was experimented in Italy. By 2006, with a turnover of 1,906 million Euros, JCDecaux Group employed almost 8,000 people in about 50 countries in the world. The company is the number one outdoor communication specialist in Europe and Asia Pacific, and number two worldwide after the American group Clearchannel. JCDecaux stands as the global leader in urban furnishing, mostly bus shelters and the like, with 318,000 advertising panels in 36 countries; as well as in advertising panels in airports, equipping 153 locations with more than 200,000 panels. In addition, it is also the number one large-billposting displayer in Europe, with 200,000 spots in 29 countries.

Over the years, innovations have continued to be introduced at all levels of the business. The company's superior ability to craft customized communication solutions for its clients led to the creation of a subsidiary called "JCDecaux Innovate", a specialized firm that offers "in-depth knowledge of innovative outdoor solutions and technologies and helps create bespoke outdoor solutions tailored to achieve brands' advertising objectives". For instance, the unit built a gigantic Louis Vuitton styled briefcase, more than four meters high, for display in the Hong Kong International Airport, while shaping a series of bus shelters in Thailand to look like Caltex gas stations. Given the growing density of city population worldwide, the more visibility the company obtains in a given populated area, the higher rate of use its display panels will have.

However, winning the licenses for advertising in large urban dwellings has proved a challenging task; it requires continuous innovation aimed at offering always more and better-perceived services to citizens and public authorities. A delicate balance has to be established between the volume of free services delivered to ultimate beneficiaries and the scope of advertising sold to commercial brands. Innovation materializes not only by permanently extending service offers delivered to cities and citizens, but also in terms of the growing attractiveness of the communication support provided to brands and consumers. The company perfectly assimilates the double nature of the targeted service beneficiaries, both local citizens and regular consumers.

Given the growing importance of safety and environmental concerns on the agenda of politicians, the company crafted several technology-based new services in order to attract the attention of city representatives. CycloCity® certainly stands as the most popular recent move of the company into the world of environmentally friendly service innovation. CycloCity®, created by JCDecaux, is a user-friendly bicycle renting service in cities and was first experimented at a limited scale in Vienna, Austria, and Cordoba, Spain, in 2003, and then expanded to Gijon, also in Spain, the following year. "*CycloCity®, It's all about service, not just a bicycle*" claims the company. It is said to be easy to use, is environment friendly and advocates healthy living. It is available 24 hours a day, seven days a week at many renting stands. CycloCity® is an ideal complement to public transport; the renting procedures are simple and affordable.

In May 2005, the service was launched in Lyons and locally perceived as some sort of an urban revolution. 40,000 local dwellers subscribed the service in the first three months, getting access to 2,000 bicycles and 175 cycle racks. After one year in operations, CycloCity® boasted the following statistics: Average journey time of 17 minutes for an average distance covered of 2.6 kilometers (1.7 miles), and turnover rate of bicycle from 12 to 15 per day. In 2007, after successfully deploying the concept in Paris within a period of five months, JCDecaux moved significantly upward and also soon became the leading bicycle renting Service Company in the world.

By understanding the importance of simultaneously addressing the distinctive needs of the three levels of receivers of the service, innovative service companies deploy efforts to balance and possibly maximize the perceived value provided to each category of players, even if their interests do not always converge with one another. Citizens don't necessarily cope well with intrusive advertisements on urban landscapes, while brand owners want highest possible return on their marketing investment. Members of multi-brand loyalty programs may want to favor a certain set of rewards that may weaken the position of some of the players in the network. Thus, a significant proportion of the value produced by these innovative service companies precisely focus on optimizing the "paying bodies" financial effort. Several examples from other parts of the world could certainly be added to this list, including the well-acclaimed Google business model based on smart advertising. However, given the particular importance in the agenda of many consumers in Europe, as well as decision makers, of issues such as regulations, public authority, social responsiveness, and others, the sensitivity to the multi-level/multi-dimensional provision of value seems well seeded amongst service companies' managing teams in the region.

Organizational Zoom . . .

ZipDial: Connecting Consumers and Brands via Disconnected Calls

In India (perhaps, this trend is also similar in many other developing countries), people use missed calls to notify the other party at zero cost. Basically, a lot of residents in the country use prepaid cellphones. Each connected call and sent text costs money; therefore, the concept of missed call was born. Say for example, if someone was to drop his friends off at their house and head home, he would call them and let it ring once or twice and hang up. That's their way of sending a signal, like "I'm home safe," without being charged for a call.

This phenomenon became a business opportunity for a Bangalore-based start-up ZipDial, established by Valerie Wagoner in early 2010. The company has just announced reaching the milestone of 400 million missed calls[21]: That's how many times people have used the company to connect with brands including Gillette, Disney, Procter & Gamble (P&G), and IndiaInfoLine, etc.

What ZipDial does is it provides advertisers and companies like Disney and KFC with a toll-free number which allows them to leverage on the "missed call" culture. If users give missed calls to a brand's number, they will get a text message in return with the latest promotion at no cost. It works like this: ZipDial issues the brand a number, which it prints on its ads, billboards, collateral and cosial media, etc., allowing consumers to call the number. The call gets automatically disconnected, so the user is not billed. The users' numbers are captured by ZipDial's platform and its business clients pull these numbers. The clients then use the mobile numbers of users who have shown interest in their product or data to send offers, coupons, or any other messages that the business wants to convey. This is important because incoming text messages are free for prepaid cellphone users.[22]

When asked what inspired Valerie and her team to start ZipDial in an interview, she said: "*We recognized two problems in the market. First, in such a diverse market, advertisers need more granular data*

[21]http://techcrunch.com/2013/03/06/zipdial-has-turned-400m-missed-calls-into-money making-connections/.
[22]http://www.zdnet.com/in/zipdial-gives-brands-marketing-opportunities-via-missed-calls-7000023564/.

on consumer preferences, and they cannot get this from internet (only 8% penetration), smartphones (only 4% penetration) or payment data (only 3% purchases on cards). Second, users get nothing personalized and advertising messages are predominantly about cricket, Bollywood, and spam. ZipDial took the typical mobile VAS model, flipped it upside down, and made the user experience simple and free for 100% of consumers (based on 'missed calls')".[23]

Essentially, ZipDial is a marketing and analytics platform that connects end users and all the major global brands. On the one hand, for end users ZipDial is of course free of charge, works on all handsets, with all operators. Consumers get near marketing personalized communication as they subscribe to the brands on their own accord. On the other hand, ZipDial works closely with their enterprise customers to design the best mobile marketing campaigns to maximize their marketing return on investment (ROI). The data that the brands get from their engagement with a large pool of consumers over mobile gives them insights which could mean a lot to marketers and advertisers, and can be used for intelligent, data-driven marketing.

To date, ZipDial has received seed funding from Mumbai Angels, Blume Ventures, US-based 500 Startups, Singapore-based Jungle Ventures, and Times Internet. While they continue to serve their Indian clients, they are now expanding internationally. In 2013, ZipDial expanded to Sri Lanka and Bangladesh, and is now setting up in Indonesia, Singapore, and the Philippines.

Indeed, there are strong trends toward the necessity to address individually expressed requests, in different layers of targeted recipients, from end consumers to business markets. Sustainable development concerns such as climate change, urbanization, demographic trends both in developed and developing economies, rising cost of energy and the shortage of vital resources such as drinking water, will increase the demand for the revamping of societal services system. It is expected that economies will particularly focus on improving the quality of life and environmental issues. In the business environment, evolution of consumer behavior, and the availability of advanced technological solutions will further lead to transformation in services to improve productivity, and generate revenues. Whatever the setting — commercial activities or social support — services address real

[23]https://sg.finance.yahoo.com/news/zipdial-receives-funding-jungle-ventures-081131098.html.

beneficiaries' issues. In general, the provision of solution over time generates continuous revenue streams.

Service Design: A Powerful Innovation Tool

For long an unread subject, service design has recently generated a great deal of attention from businesses as well as a significant number of academic literatures. Frequently defined as "interaction design", aiming at enhancing the recipient's experience, service design now centers around the concept of building trust at acceptable cost, as Michael Wescott, from the Design Management Institute, puts it, "whether it's a well-designed customer service experience, a more effective connection with a doctor, a well-spent moment on a smart phone, a more inspiring educational experience, a more engaging or efficient use of a toaster, or an entertaining travel experience".[24]

Particular attention is given to "touch-points" through which companies interact with service beneficiaries. Researchers distinguish between designable and non-designable touch-points and recognize three main categories of these: The designable physical environment and tangible commodities, the equally designable virtual interface such as ATMs, websites or mobile apps; and finally the sometimes designable, sometimes non-designable service personnel with their skills, attitude, and knowledge.[25] Kyongsill *et al.* further classified designable touch-points by design discipline: Space design, product design, identity design, advertising and graphic design, and new media design. They advocate for an orchestration of designable touch-point in view of altering business strategy by manipulating key factors, from passive interaction to active, or from utilitarian motive to hedonic, for instance.

For Marc Stickdorn and Jakob Schneider, service design is "user-centered" since customer participation is necessary for the delivery and often for the design itself.[26] It has to be "co-creative" involving not only

---

[24]M Wescott (2013) The changing nature of service and experience design. *DMI review* 24(3). Fall.

[25]Kyongsill Lee, Kyung-won Chung and Ki-Young Nam (2013) Orchestrating designable touchpoints for service businesses. *DMI review* 24(3). *Op. Cit.*

[26]M Stickdorn and J Schneider (Ed.) (2011) *This is Service Design Thinking.* Amsterdam: BIS Publishers, Wiley.

the beneficiaries but also all the various stakeholders in defining the service proposition. Being dynamic processes, services are subjected to crucial time management. So service design is "sequencing" and also "evidencing" since intangible value delivery should be made somehow visible to the recipient. In addition, a "holistic vision" proves necessary to the successful shaping of any new service experience.

In their book defending that every work is a theater, B. Joseph Pine II and James H. Gilmore explore how successful companies using services as the stage create experiences that engage customers in an inherently personal way. Following that thought, companies have to determine what form of theater to perform and how to stage customer perception of any service experience.[27] Also, since services are delivered through multiple distribution channels that are overlapping, service design now addresses the challenge of shaping multi-channel service experience. Satu Miettinen and Essi Kuure express the importance of service design in the co-creation of value, mobilizing different service design methods and a wider approach that integrates service thinking and understand how the user interacts with service rationales and relations.[28]

Design-driven service innovation leads to the assessment and reinvention of the entire customer journey, from pre to post consumption. Ultimately, a well-designed service experience certainly enhances the quality of life of the recipient. However, as Henry Chesbrough mentions, there is a deep tension between scalability, which makes service innovation efficient, and customization, which makes service innovation attractive to customers.[29] Managing this tension using unique ways definitely stands as the most critical part of any service design attempt.

As a summary, a three-stage process of service design for innovation is suggested, which includes: Ideation, design, and delivery set of operations (Figure 3.2).

The ideation phase may significantly benefit from the implementation of the open service innovation framework, given the necessity to involve recipients and providers in the process to generate advanced solutions. As a co-creating user-centered interface, the design phase requires the

---

[27]BJ Pine II and JH Gilmore (1999) *The Experience Economy, Work is Theater and Every Business a Stage*. Boston, MA: Harvard Business School Press.
[28]Satu Miettinen and Essi Kuure (2013) Designing a multi-channel service experience. *DMI review* 24(3). *Op. Cit.*
[29]H Chesbrough (2011). *Op. Cit.*

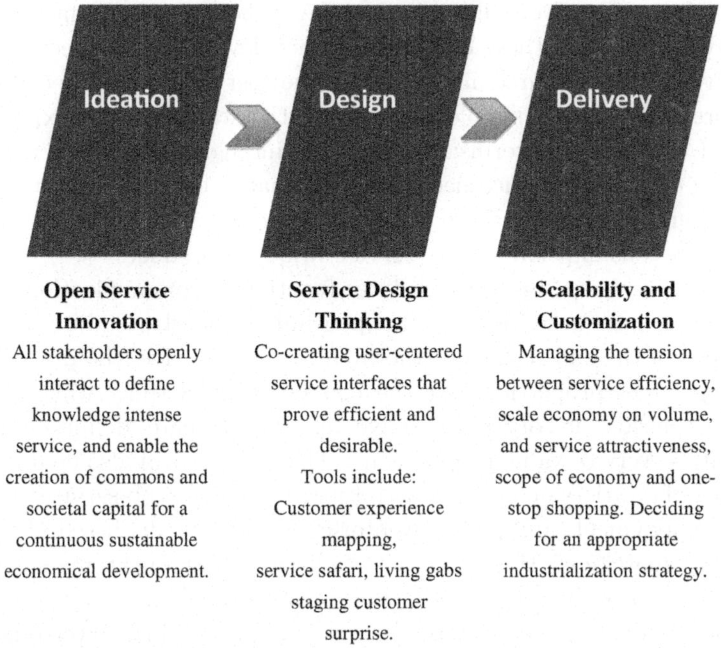

Figure 3.2: Key Stages of Service Design for Innovation.

use of "customer experience mapping", "service safari" or "living Labs" tools to secure customer surprise and enhance the perception of the received value. The delivery phase of the design process revolves around the necessity to address the tension between service efficiency and service attractiveness, leading to the adoption of an appropriate industrialization strategy.

Assembling the Most Efficient Set of Physical Support and Service Provision

Service innovation should be formulated in parallel with design and development of the tangible product, or physical support, which may allow for its delivery and consumption. Anticipating the "service needs" of the tangible support leads to the consideration of how the physical characteristics and parameters of the product match service objectives and constraints, just as designing the object itself leads to expressing specific needs in terms of technical considerations. Consequently, managing the product during its lifetime, in terms of service needs, is impossible to untangle from managing the lifetime of the tangible product. A coherent management of the interactions between these different product development aspects requires the consideration of three dimensions of the "service value proposition" or real "product" in more simple terms: First, the "tangibles", as expressed in physical/technological terms; second, the "intangibles mix", as expressed in terms of market service needs and the match between both these and the customer needs over time; the third "product" dimension refers to the customer's consumption of the "service value" over time.

Aggravated by insufficient inter-functional communication or co-ordination, these activities may happen independently, leading to lack of coherence of the end result. Furthermore, to be effective, the process must match product properties over the course of its lifetime, with changing customer needs, in order to arrive at the ultimate definition of the company's "service value proposition". This process will also enable in the planning of a variety of ancillary activities: Additional web-services, wireless electronic payment, and parts inventories, etc. As we discussed above, the iterative "matching" that occurs as "physical component managed over time" and "service component managed over time" gradually coalescing into the "ultimate value proposition", is a spectrum of the inter-functional conflicts that exist in most firms. Firms that recognize the need to bring a multi-functional team to the design process are able to move along a direct path to the end result.

The remaining question deals with organizational structure. Considering that some functional separation is necessary for developing professionalism and the quality of the work carried out, but also considering integration to be a condition for ensuring coherence in managing the product throughout its lifetime, as well as for controlling the total cost of its development and its consumption, attention was concentrated on the identification, the categorization and the establishment of integrative units in the service organization. The concepts of integration and separation, permanently linked together, are richer than the pair of centralization and decentralization, even though the latter forms the basis for delegation of functional responsibilities within the service organization. Co-ordination is required in virtually every activity that comprises a service function. Thus, as an example, modifying a replacement part should lead, theoretically, to adapting the documentation identifying it, to modifying the kits and updating stock, to transmitting to the replacement logistics function the anticipated rate of availability, as well as to managing the circulation of these parts and rebuilding the former models to replace them, etc. For a long time, most after-sales managers would have admitted that such serially-connected actions are difficult to carry out because of the absence, within their organizations, of integrative units explicitly responsible for guaranteeing this sequence of activities.

Chapter 4

COMPETING THROUGH SUPERIOR PRODUCT DESIGN

The purpose of innovation is to deliver enduring value which is frequently, but not always, expressed through the deployment of new technologies and the resolution of complex problems. If no significant improvement is brought about, in society in general or at the company level in particular, then it can be said that there has not been any innovation. According to Peter Drucker, innovation is considered one of the two key valuable functions of any business, together with marketing. Everything else, in his opinion, only serves to consume resources instead of creating value for the company. In his opinion, continuous and permanent innovation requires a willingness to consider change as an opportunity, and only *per se*. David Ricardo introduced the concepts of competitive and comparative advantage as prerequisites for financial exchanges, thus opening the path for a permanent effort to strengthen that advantage through innovation.

Product innovation has gone through different stages, beginning with the product-oriented or technology-pushed stage. In the post-World War II era, Americans were coming off wartime shortages and were in the mood to buy huge amounts of goods that manufacturers produced. The dominant mindset at that time was to consider the market as the receptacle for products that companies delivered. Engineers were more product-oriented than consumer-oriented, designing new products that might or might not find win consumers' hearts and minds. Kim B. Clark and Steven C. Wheelwright from Harvard introduced product development as a central focus of competition in the 90s.[1] They identified the forces that were driven by the growing importance of product development, i.e., changes in competition, customer demands, and technology; and explained how these forces had created a competitive imperative for speed, efficiency, and high quality in the development process.

Since the mid-80s, a worldwide change in concept has been observed in an effort to obtain new sources of competitive advantage based on a wider view of the service content irrespective of the type of offering to the market. Faced with declining margins in many industries, in manufacturing activity, it has proved essential, in a first phase of evolution, to find new ways of reconstituting profits by providing a new concept of advanced customer service in addition to the sale of tangible products, eventually proving an integrated part of the product itself. As explained earlier, the emerging experience economy has provided organizations with a new paradigm in relation to the notion of value delivered to the market. On the one hand, in many industries, the true nature of consumption switched from the acquisition of tangible products to the use of functionalities and utilities. On the other hand, in many traditional "service" businesses, the provision of more sophisticated physical support has been introduced as an integral component of the "products" delivered to the markets.

Entering the 21st century, companies are facing intensive and increased competition on a global scale. More competitors are capable of world-class performance. To build adaptive capabilities into their organizations, companies recognize that innovation is a complex process that requires sound investment in research and development, as well as significant marketing expertise that focuses on satisfying consumers' wants and needs. New products are more broadly defined as any product, service, or ideas not currently made or marketed by a company, or which may be perceived as new by customers. Instead of the mono-approach, in which technology or markets drive innovation, new product development now requires an integration of technology, marketing, product design, engineering, and manufacturing capabilities. We will discuss a new framework of three dimensions in the following sections. Henderson and Clark argue that innovation rarely involves dealing simply with a single technology or market but rather a bundle of *knowledge* that is brought together in a

[1] KB Clark and SC Wheelwright (1993) *Managing New Product and Process Development.* New York: The Free Press-Macmillan.

configuration.[2] Successful innovation requires companies not only to get hold of the knowledge of the technological components but also their arrangement, referred to as the architecture of innovation.

ASSESSING THE CHALLENGES OF PRODUCT DESIGN

As competition escalated during the 80s and consumers became more skeptical and selective about the types of products they purchased, companies found it more and more difficult to rely on persuasive sales techniques to sell products. Retailers were increasingly worried and concerned when these products did not move off shelves as quickly as expected. Companies started to realize that they had to know more about their target markets. What were the wants and needs of the people and who were buying their products? How could they satisfy these wants and needs?[3]

This stage of evolution was brought about by the emergence of the market as the driver of innovation. Instead of being technology-driven, new product development evolved into a market-led process in which new products emerged from well-researched customer needs. The new product development process was placed in the hands of marketers who knew consumers' wants and needs. However, as explained in the early 80s by MIT specialists in an iconic book,[4] marketing and R&D should not be mutually exclusive alternatives, since a successful new product strategy requires an effective integration of the two functions. Nevertheless, customer demand became the pilling force of the product through its development process.

By mid-90s, the characteristics of successful product development, as expressed by another team of MIT scholars, were well understood.[5] The performance of a product development effort was then commonly assessed

---

[2]RM Henderson and KB Clark (1990) Architectural innovation: The reconfiguration of existing product technologies and the failure of established firms. *Administrative Science Quarterly* 35(1), pp. 9–30.

[3]http://www.referenceforbusiness.com/management/Mar-No/New-Product-Development.html.

[4]GL Urban and JR Hauser (1980) *Design and Marketing of New Products.* Upper Saddle River, New Jersey: Prentice Hall.

[5]KT Ulrich and SD Eppinger (1995) *Product Design and Development.* New York: McGraw-Hill.

through five specific dimensions:

- **Product quality**, defined in a holistic view as meeting the market expectations.
- **Product cost**, determining future profit and including manufacturing investment.
- **Development time**, reflecting how responsive the firm can be to competitive forces and technological advancements.
- **Development cost**, as a fraction of the company's investments to generate profits.
- **Development capability**, as an intangible asset that the firm can use to develop products more effectively and economically in the future.

Ulrich and Eppinger also identified some of the characteristics that make product development challenging. Managing trade-offs in a way that maximizes the success of a product, for instance, is one of the most difficult aspects of the process. Coping with the dynamics of technology improvements, evolution of customer preference, and competition intensification are formidable tasks. Thousands of detailed decisions have to be made while developing a product of modest complexity, raising new challenges. Finally, product development decisions, due to time constraints have to be made quickly, with or without complete information.[6]

Even if hardly communicated by companies, developing new products frequently fails for all these reasons. A series of methodologies have been designed to actually reduce the risk of ending up with unsatisfying developments. Dassault Systèmes group of France has developed a series of successful advanced software solutions such as Catia (considered as a worldwide leader in 3D computer-aided design (CAD)), Solidworks and Exaleads.[7] Other popular tools include PTC series for Computer Aided Design (CAD), Product Lifecycle Management (PLM), Application Lifecycle Management (ALM), or Service Lifecycle Management (SLM).

Organizational Zoom . . .

PTC Solutions and Schindler

Founded in 1985, PTC in Needham Massachusetts serves today more than 28,000 customers. It is of the view that manufacturing companies can simplify the complexities of global product development through

[6] *ibid.*
[7] http://www.3ds.com. Dassault Systèmes 2015 company website.

> efficient team collaboration. PTC has streamlined the use of strategic supply chain partners.[8]
>
> By implementing common, standardized tools and processes, team productivity has increased, management of local market requirements has improved, and access to local development resources has resulted in lower cost for the company. Implementation efforts have resulted in faster time-to-market reduced regulatory compliance risk and increased geographic revenue diversity.
>
> Bernhard Gysi, head of R&D, services and testing at Schindler once said: "Without process standardization, we would have needed to double or triple the human resources at our overseas sites. Accordingly, we have realized a 200–300% improvement of efficiency in our product development".

Traditionally, the upstream side of the development process, when applied to a physical product such as an automobile or a computer, is made up of five phases: Concept development, system-level design, detail design, testing and refinement, and production ramp-up. Each of these phases requires the involvement of numerous departments such as marketing, design, system engineering, manufacturing, finance or purchasing. During the concept development phase, the needs of the target market are identified, alternative product concepts are generated and competitively evaluated, and finally a single concept is selected for further development. Looking at the downstream side of the process, we may add important tasks such as outbound logistics, after-sales support, and recycling.

In the 21st century experience economy, the gap between manufacturing activities and service operations has clearly narrowed. Manufacturing increasingly seems to follow the lead of the service sector wherein customer service is becoming more important day by day and tangible products are often tailored to the needs of customers, through "mass customization" or "late differentiation" production systems. The general growing impact of customer service and the generalization of smart devices in particular, do change user and consumer expectations of product manufacturers.[9] Also, the role of the industrial design has always been to humanize the new technologies by improving the experience of using them. But the more dematerialization accelerates, the more this role grows in importance.[10]

---

[8]www.ptc.com. PTC 2015 company website.

[9]J Heapy (2011) Creating value beyond the product through services. *DMI Review* 22(4), pp. 15–21.

[10]C Roux (2011) The new meaning of product design? *DMI Review* 22(4).

On the other side, direct contact between the provider and the customer may be unnecessary in many service activities; and some services, at least half-produced services, can actually be held as inventories and traded internationally. Service response logistics, as a part of the customer service function, involves activities such as taking and recording customer orders, scheduling personnel and equipment for service calls, and executing service calls in an efficient way. Sophisticated information and communications systems are often used to assist in these processes. Service excellence typically emerged as a large component of the marketing mix with commensurate focus placed on it by management, either in manufacturing or in typical service organizations. Service excellence also qualifies today as a key and integrated component of the product itself.

As a result, in contemporary context, the value associated with any given product, being a necessary combination of physical and service substances has to be appraised following multiple dimensions. Two dimensions are traditionally well documented in each and every attempt to develop new product, especially in the manufacturing context: *Functionality* and *pricing*. Functionality, however, has to be approached in its broader modern significance. Functionality is a notion that expand significantly above the sole notion of adequate technical response to usage requirement. Under the Service/Product hybrid paradigm, it relates also to the accessibility of the offering, the "user friendliness" of the design, the simplicity and safety in the use or consumption, as well as the clarity of the associated payment system. Price definition relates to the notion of perceived value for the beneficiary and to the total costs of access and consumption including hidden costs. While getting better ranked on the price scale might usually mean reducing the cost in absolute term; significantly improving the perception of acquired value associated with the consumption of the offer unveils a more attractive value equation for both the provider and its client, even if the actual price rises.

Given the growing importance of brand equity and reputation of companies or any given product, we add an "Image" criterion to the traditional paradigm that would clearly be disconnected to the fourth dimension which we have picked in the notion of "Emotion". Some products/services might enjoy a strong and somehow positive image, though their mention may generate mostly negative emotions, e.g., like hospital and recovery health care or sanitary inspection, among others. Creating emotion has become a key feature expected by the recipient of almost any type of value proposition. Emotion should emerge while accession, consuming or acquiring a "product". The emerging importance of emotion as an objective in "product" design has mostly been introduced by the advocates of the "Design Thinking" philosophy and methods. Verganti, insists on the difference between "design-driven" innovation and the better established technology- and marketing-driven innovation strategies.[11] Through the "design-driven" approach, innovators are precisely working to create emotions and communicate sense and values while defining a new product. The notion echoes with the conceptual framework suggested by Tom Kelley and Jonathan Littman,[12] and then by Tim Brown when leading the IDEO design firm in California.[13] Roger Martin, Rotman School of the University of Toronto, has developed series of applications with the "design-thinking" approach to tackle multiple corporate situations and challenges that argue for greater creativity and emotional content in managerial decisions.[14]

Based on the four multi-dimensional user appreciations of any product/service package described above, we are suggesting a simple evaluation chart aiming at facilitating development decisions. Introduced in Figure 4.1 and called the "blooming spider" method, the chart contributes to state the position of any existing or projected offering following a four scaled lines of performance: Function, price, image, and emotion. The so-called "perfect offer" is reached when the product/service delivers "delight" to the user; being a great value for the resource invested by the beneficiary, and enjoys an iconic image instilling an inspiring as well as rewarding emotion.

EFFECTIVE DESIGN FOR CROSS-DIMENSIONAL INTEGRATION

Increasing competition and more demanding customers suggest that firms which do not focus on new service development have the risk of falling behind. At the same time, accelerating technological developments are also opening up remarkable opportunities for developing innovative services. A set of tangible products, a mix of intangible services associated with them, and the management of the relationship of the two over time, as customers' needs and the

[11] R Verganti (2009) *Design Driven Innovation: Changing the Rules of Competition by Radically Innovating What Things Mean.* Boston, MA: Harvard Business Press.
[12] T Kelley and J Littman (2005) *The Ten Faces of Innovation: IDEO's Strategy for Defeating the Devil's Advocate and Driving Creativity throughout Your Organization.* New York: Currency Doubleday.
[13] T Brown (2009) *Change by Design: How Design Thinking Transforms Organizations and Inspires Innovation.* New York: Harper Collins.
[14] RL Martin (2009) *The Design of Business: Why Design Thinking is the Next Competitive Advantage.* Boston, MA: Harvard Business Press.

Figure 4.1: The Four Key Components of Product Perceived Value.
Source: Adapted from H Mathe (2012) Leading in service innovation: Three perspectives on consumer service value delivery. In P Wieser, F-L Perret and C Jaffeux (eds.) *Essentials of Logistics & Management*. Boca Raton and Lausanne: EPFL Press and CRC Press, Taylor & Francis Group.

environment in which the product is used change, constitute the three dimensions that together fully define the complete "enhanced value proposition" of any company. To be successful in this endeavor, the company's management should understand that service, even in the so-called "manufacturing" environment, is always a corporate-wide activity that requires inter-functional coordination in both life cycle analysis and in carrying out the myriad of necessary activities. Defining a robust service innovation requires the integration of the service dimension as a component of the company's offer to the market in each life cycle stage of the beneficiary–provider relationship. Finally, technology provides the resources needed to actually put the "enhanced value proposition" concept into practice.

The distinction between the product as a system and the product as a set of components highlights the idea that successful product development deploys two types of *knowledge*. First, it necessitates component knowledge, or knowledge about the core design concepts and the way in which they are implemented in a particular component. Second, it requires architectural knowledge or knowledge about the ways in which the components are integrated and linked together into a coherent whole.[15] Radical innovation

[15]RM Henderson and KB Clark (1990). *Op. Cit.*

establishes a new dominant design and, hence, a new set of core design concepts are embodied in components that are linked together in a new architecture. Incremental innovation refines and extends an established design. Radical and incremental innovations are two extremes along the two dimensions of component design and the system of linkages between core concepts.

Further on, we provide a new framework of product development featuring three key dimensions: The "tangibles", as expressed in physical/technological terms; the "intangibles", as expressed in terms of market service needs; and the match between both of these and the customer needs over time. These three dimensions should be simultaneously considered during new product development process, as they are mutually correlated and reinforce each other. The physical product should be designed so that it is adapted to service objectives and constraints; while the service package should be designed to go along with the tangible product. Moreover, defining the mix of services to be provided requires the understanding of customer needs throughout the entire lifecycle of the customer–supplier relationship. So, a number of relationships may be established between the different activities that occur during the "product's lifetime".

Coherent management of the interactions between these different product development aspects considers three dimensions of the "holistic value proposition" or real "product" in more simple terms:

- The "tangibles", as expressed in physical/technological terms;
- The "intangibles", as expressed in terms of provided service experience;
- The match between both the "tangibles" and the "intangibles", and the customer needs over time. It refers to the firm's ability to manage the customer's relationship over time.

Figure 4.2 suggests a representation of the model through which the three dimensions of "product" development can be managed.

Specifically, service innovation should be formulated in parallel with design and development of the tangible product, or physical plane, which may allow for its delivery and consumption. Anticipating the "service needs" that can be embedded in the physical product itself is key, such as remanufacturing ability on self-diagnosis device. The effort leads to an assessment of how well the tangible characteristics of the product match service objectives and constraints, just as designing the object itself leads to expressing specific needs in terms of technical considerations. Consequently, managing the product during its lifetime, in terms of service needs, is impossible to untangle from managing the lifetime of the tangible product.

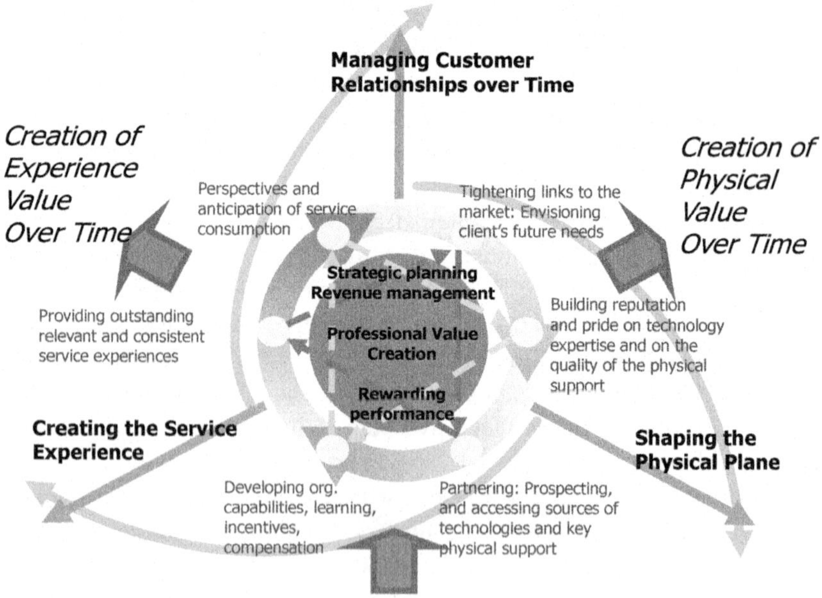

Figure 4.2: Managing the Three Dimensions of Product Development.

Shaping the Physical Plane

As opposed to the traditional view of centering any product on the supply of a tangible item, we advocate for the offer to be primarily considered as a delivered service, whatever be the administrative nature of the commercial transaction: Regular sale, renting or leasing. It has been recognized for a long time that physical products or attributes are an indispensable dimension when creating the conditions for a service experience.[16] According to Bitner, physical/technical enablers include physical signs, symbols, products, and the infrastructure. These elements are often related to the multi sensorial experiences, such as visual images, tactile impressions and sounds, the generation of individual experiences, imagination, and the emotive aspects of the service usage experiences.[17] Services are intangible; nevertheless, they are often released through the use of physical products.[18]

---

[16]MJ Bitner (1992) Servicescapes: The impact of physical surroundings on customers and employees. *Journal of Marketing* 56(2), pp. 57–71.
[17]EC Hirshman and MB Holbrook (1982) Hedonic consumption: Emerging concepts, methods and propositions. *Journal of Marketing* 46(3), pp. 92–101.
[18]LG Shostack (1977) Breaking free from product marketing. *Journal of Marketing* 41(2), pp. 73–80.

A characteristic of technology-based services is the fact that they are accessed through some kind of "service access equipment", e.g., a mobile phone, a pager, or an ATM, etc. The access equipment is thus a physical piece of hardware that is necessary in order to consume the service. In addition to the access equipment, other physical/technical enablers that aid in providing the service include the underlying systems through which the services are achieved in collaboration with the more physical products. For example, the mobile telephony network is a prerequisite for the offered service or value proposition. The coverage and capacity of the network will decide where the service can be used, how many people can use it at the same time, and how fast it will perform. The physical access device and the technical infrastructure are in a kind of symbiosis, both dependent on each other's existence.

When developing a new product, engineers and designers should always consider the actual value propositions made to the customer, i.e., the functionality, image, and emotion. The functionality of a value proposition includes what is possible to do using the physical/technical enablers available (e.g., the user can make calls or take photos using his mobile phone). The image and emotion dimensions provide the intangible features. Mental images are usually physical symbols on a certain product in the form of a logo that gives the users a mental image, i.e., an indication of what to expect from a service offering. The emotional value proposition, including brand reputation and themes, is used to engineer the intended, favorable service experience and is viewed as an expression of corporate culture and strategy. Intangible dimensions can also reflect the specific norms and values of a company, e.g., its codes of conduct and corporate narratives (Edvardsson and Enquist, 2002).[19] The tangible and intangible dimensions of a product together contribute to the service experience and user value.

Organizational Zoom . . .

Apple: Developing an Ecosystem that People don't want to leave

While Apple's new product introductions, particularly the iPhone and iPad, have received the most attention, iTunes too has expanded

[19]B Edvardsson and B Enquist (2002) Service culture and service strategy: The IKEA saga. *The Service Industries Journal* 22(4), pp. 153–186.

dramatically over the past decade. Efforts by Amazon, Google, Microsoft, Samsung and others to rival iTunes have done nothing to stop the phenomenal growth of Apple's software and services.

With the introduction of iTunes, Apple has entered a new market and challenged existing business models. This is a key element of Apple's approach to innovation. While many see the product as the focus, for Apple it is the deep integration of the product and the service that allows it to fundamentally change the underlying business model operating in the sector. First it was music and now it is movies. *Via* its iTunes service, Apple has challenged the way the music industry traditionally operates, and the company has expanded its music videos, audio-books, television shows, movies, and podcasts repertoire still further with major links to all the key studios. In 2007, iTunes had its three billionth download and this helped music related products and services account for a significant \$2.5 billion of sales. Of course, customers also expect a suitable "Apple" experience when they enter a store. And they get it. Apple now has numerous stores worldwide and this has redefined computer retail.

All of the above innovations have translated into corresponding financial success. Market capitalization as of January 2015 stands at \$640.7 billion and total sales in 2014 increased to \$182.80 billion, while gross profit reached to a staggering \$70.5 billion. These are ample evidences to highlight that innovation in Apple is not hype: Apple really delivers its goal to provide "innovative integrated digital lifestyle solutions" and is a company that consistently demonstrates the power of innovation in driving profitable, sustained growth.[20]

Clearly, Apple is building a stronghold around its suite of product/services/experiences and it proves almost impossible to break Apple's monopoly. iTunes itself is more than a service; it's a platform upon which subsequent hit products, including the iPhone and the iPad, build themselves. Every single Apple product complements and completes each other. They have created desirable offerings and features to create consumer lock-in and switching costs. Get an iPod, and you can download content *via* iTunes, and access them through iCloud, which allows consumers to seamlessly share media across devices such as the iPod, Apple TV, and the Mac PC. Apple TV's Airplay allows music or videos from the iPhone or iPad to be effortlessly shared over WiFi. This sort of control over the entire user experience or process, from hardware to software, strengthens customer loyalty and

[20] http://innovationleaders.org/apple_company_profile.html.

provides lock-in. Indeed, it is Apple's integrated ecosystem of products and services which makes it hard for customers to leave once they are in. In fact, 21% of iOS consumers say they'll never leave Apple's ecosystem.[21]

Creating the Service Experience

As competition increases and technological advances accelerate the renewal of tangible products, companies face mounting pressure to develop innovative products and services. For the last ten years and more, customer service has already proved to be a significant source of competitive advantage for both manufacturing companies and retailing firms. While successful manufacturing companies have built on their core capabilities, they have moved beyond the factory gate to tap into the valuable economic activity that occurs throughout the entire product life cycle.[22] Furthermore, in many traditional "service organizations", attention to customers or clients has dramatically increased, enabling leaders to remain ahead of the pack or conversely, provide the added thrust to allow a challenger to attack and overtake the leader. Currently, service is definitely reinforcing its preeminence as a key activity for manufacturing companies operating in economically developed countries, and new service firms are flourishing everywhere.

In the contemporary business world, pioneers are moving toward a dynamic exchange relationship perspective that involves performing processes and exchanging skills and services in which value is co-created with the customer.[23] Previously, value was thought to be embedded in the product itself. Now a new value perspective that emphasizes "value in use" is becoming dominant,[24] where value is realized only when a service is used. The shift from selling pure products to developing value-added services has begun across different industries.

---

[21] http://www.forbes.com/sites/jump/2012/11/12/take-a-lesson-from-apple-a-strategy-to-keep-customers-in-your-ecosystem/.

[22] R Wise and P Baumgartner (1999) Go downstream: The new profit imperative in manufacturing. *Harvard Business Review*.

[23] CK Prahalad and V Ramaswamy (2004) *The Future of Competition: Co-creating Unique Value with Customers*. Boston, MA: Harvard Business School Press.

[24] SL Vargo and RF Lusch (2004) Evolving to a new dominant logic of marketing. *Journal of Marketing* 68(1), pp. 1–17.

Organizational Zoom . . .

Rolls Royce: Power by the Hour

The service concept has been elevated to great heights by Rolls Royce, the aero turbine manufacturer. Not only do commercial customers demand round-the-clock, round-the-globe support for their power units, the military does not even own its units! Rolls Royce holds ownership of all cores and spares. In fact, the company first started their "power by the hour" program over 20 years ago, to deliver "Total Care" that improved their customers operation. Over time, they introduced sophisticated maintenance contracts which guaranteed engine availability. Customers pay for engine power as they use it. Rolls Royce provides repair and support services for engine components around the world, and charges for usage as agreed with the customer. It is in the firm's interest to maximize reuse capability, subject to other client constraints. The client is offloaded from maintenance tasks and has a clear view of his future liabilities.

Michelin Fleet Solutions

Recognizing that tires are service intensive products, Michelin developed Fleet Solutions in the year 2000 as a new division. The idea behind this move was to provide unique value to customers by taking charge of tire management in order to achieve full potential of Michelin tires. They no longer just sell tires, but sell kilometer instead, providing customers with the ability to drive. Customers pay a monthly fee depending on vehicle use (number of kilometers driven).

Pfizer: Healthcare beyond the Pill

Biopharma companies are going beyond the pills. Pfizer has started a "Lipitor for You" program, which includes the $4 co-pay card, the option to receive direct delivery of the prescription, and periodic emails with health information and reminders to refill Lipitor prescriptions, to encourage continued use of branded Lipitor after patent loss.[25]

Customers or end users are thus both the co-creators and the judges of service value, and this is where service design comes in. Service design can yield both financial rewards and satisfied consumers by helping companies to reach an in-depth understanding of its customers, and shape a service that meets their real, often unrecognized needs. Service design involves

[25] http://online.wsj.com/articles/SB10001424052970204528204577011492595048250.

planning and organizing people, infrastructure, communication, and material components of a service in order to improve its quality and the interaction between service provider and customers.[26] The purpose of service design methodologies is to understand the behaviors and motivations of the customers, and design according to their needs, so that the service is user-friendly, competitive, and relevant to the customers. Within service design, service interfaces are designed for intangible products that are, from the customer's point of view, useful, and desirable, while they are effective, efficient, and different for the provider. Service designers draw on methodologies from fields such as ethnography and journalism to gather customer insights through interviews and by shadowing service users. They watch and interpret needs and behaviors and transform them into potential future services. Many observations are synthesized to generate concepts and ideas that are typically portrayed visually, for example, in sketches or service prototypes. Service design may inform changes to an existing service or create a new service.

Services can contribute to increased demand for a tangible product and, as a result, to higher market share for the firm. Support services, including delivery, influence purchasing behavior at the initial acquisition, and again later when the customer, needing to replace the equipment, decides to reinvest in the company's product. Nevertheless, it is worth noting that in most advanced markets, the more primitive concept of "technical support service" has increasingly been overshadowed by customers' expectations for a product's availability both in time and space, its functions are operational when and where needed, and at a guaranteed cost. Hence, in the eyes of the final customer, acquisition cost remains the determining factor for the attractiveness of both packaged consumer goods and most simple industrial equipment. However, the services offered by the manufacturer can subsequently add considerable value for the distributor or retailer. For medium and heavy equipment, greater range and quality of services become the more dominant discriminating factors, with the customer's dual objective being to maximize the operational availability of the equipment, and to guarantee its adaptability over time to changes in use. Delivery and support services, whether or not provided by the manufacturer and deemed to be insufficient, may open the door to new competitors who are evaluated, above and beyond the price and performance characteristics of

[26]http://www.service-design-network.org/intro/.

their product, on the quality and range of the provided services, before and after sales.

Distribution and support service functions, aimed at reinforcing customer loyalty and profitability in durable goods industries, are supposed to accomplish many tasks:

- To deliver and implement the product? Certainly, but not only.
- To fix equipment efficiently? Yes, but not only.
- To provide appropriate parts at the right place and in an appropriate amount of time? Yes, but not only.
- To train the customer to most effectively use the product and to perform self-maintenance? Yes, but not only.
- To design the product in order to guarantee a low life-cycle cost in terms of both acquisition and usage, but also to facilitate recycling of materials and components? Yes, but not only.

The "Strategic Service Mix" methodology, as a way to provide value to the customer, may be defined as the correct balance of different individual services produced before, during, and after the sale of any product, to meet user needs over the lifetime of the customer–provider relationship. Product "design to cost" requires defining the characteristics and individual costs of all major functions: Production, physical distribution, maintenance, and development, over the total life of its use. But, in addition, the primary objective of the Service Mix Plan is to determine and then to ensure the various activities required during the following three stages:

- **"Preparation for use"** includes all the activities that allow the customer to obtain the tangible product (e.g., physical distribution), its adaptation to the workplace (e.g., installation), and then its startup. Less visible but perhaps more important, prior to these activities that occur just before usage begins, are the myriad of preparatory activities undertaken much earlier. These activities accompany design so as to ensure that the tangible product really meets the customer needs throughout the period of usage.
- **"Information and training"** represents the vital function of collecting and circulating technical and economic data, including designing and distributing technical manuals for instruction and maintenance. Services such as demonstrations and on-site training play a major role among these activities. Customers are always looking for more information about the handling of their orders. That is particularly true for the

express transportation sector where providers are largely competing on the capabilities of their information systems.

- **"Maintenance of durable equipment"** refers to a number of activities that take place after installation: Repair, reconditioning, and re-manufacturing. Too often however, these tasks are not always carried out effectively because they are not coordinated with other functional activities. Material recycling that prolongs the process of managing the tangible product throughout its useful life is also becoming a significant activity.

In short, service mix strategies can represent a way for the manufacturing or retailing company to increase its competitiveness, often considerably. After identifying the impact that the service mix has on the intensity of rivalry between competitors, using overall product performance measurements or a coherent combination of tangible goods and service operations, the impact of service can be ascertained as it relates to the industry for the tangible product, as well as relative to stand-alone service competitors. Support of the physical "product" plays a profitable role in the effort to build barriers to the entry of new competitors by increasing switching costs. The development of substitutes to the manufacturer's after-sales support, such as "do-it yourself maintenance" or the expansion of networks that specialize by physical subcomponents in some industries, tends to markedly modify the rules for competition through service.

Organizational Zoom . . .

Philips: Lighting as a Service

Washington, D.C. is following in the footsteps of a number of cities that are or already have switched to LED lights. However, unlike the other cities paying high costs of making the shift upfront, Washington is taking advantage of the first turnkey "lighting-as-a-service" model that Philips Lighting (which manufactures the LEDs) offers.

While the model is commonly used to finance building retrofits, it hasn't been used for lighting. The city will upgrade more than 13,000 lighting fixtures to a custom-designed LED lighting solution in all its parking garages. Philips will pay the upfront costs of installation and provide a 10-year maintenance contract. Philips will be compensated through the estimated $2 million in energy savings the retrofit produces each year. The project starts in 2014 spring and will take about a year.[27]

[27] http://www.wmata.com/about_metro/news/PressReleaseDetail.cfm?ReleaseID=5613.

Under the 10-year contract, outdated and inefficient lighting will be replaced with new high-efficiency LED lighting that will create a brighter environment for riders, use less electricity, and lower costs. The bulbs will automatically adjust based on the amount of available natural light. Designed to each garage's unique characteristics, the new LED lighting will direct light when and where it is needed. Each fixture will respond to its environment by automatically dimming, in response to motion and ambient light from the sun or other sources.

In addition to the energy savings, Philips will provide the maintenance of the lighting system for the life of the contract, and software will alert Philips when repairs are needed. This contract award reflects Philips' commitment to improving the system's infrastructure for the customers and citizens for years to come by, providing not only the lighting bulbs but more importantly continuous services, hence maintaining a long-term relationship with Washington Metro.

"With digital lighting systems, we really need to break with conventional thinking and look to the services and delivery models of the software industry to understand the future of lighting and how we can remove one of the greatest barriers to adoption: The upfront costs," Bruno Biasiotta, President and CEO of Philips Lighting Americas, said in a statement.[28]

The push for LED lighting

The Washington Metro turned to Philips because it wanted to brighten up its parking garages while meeting its sustainability goals. The switch to LEDs will reduce energy costs at the garages by 68%.[29] A survey of 300 mayors released by the Mayors Climate Protection Center showed that switching to LEDs is a high priority (82%), second only to public building retrofits (86%). LEDs and other energy-efficient lighting are the "most promising technologies" for reducing energy use and carbon emissions, they said. The most significant barriers are budget constraints (84%) and upfront cost (71%).[30] Like many cash-strapped organizations, it did not have the upfront capital, despite the energy and maintenance savings. The savings from the maintenance alone was expected to be about $600,000 annually.

In fact, one of the drivers for LEDs in parking garages and street lights is the quality of light and the ability for networking. The Philips Luxeon LEDs will be networked *via* ZigBee and will adjust to daylight

[28]http://www.greenbiz.com/blog/2014/01/27/philips-lighting-teams-mayors-power-led-use-cities.

[29]http://www.greentechmedia.com/articles/read/washington-metro-will-install-leds-at-zero-cost.

[30]http://www.greentechmedia.com/articles/read/washington-metro-will-install-leds-at-zero-cost.

and learn the patterns of each garage to adjust when they operate, as well as alert the maintenance team if a bulb burns out. At a meeting of the US Conference of Mayors in 2013, they adopted a policy that affirms the importance and benefits of advanced lighting systems in cities. At the January meeting in 2014, they formed a partnership with Philips Lighting to help mayors with this effort. Through the Mayors' Lighting Partnership, Philips will provide free energy audits, technical assistance and its new financing option.[31]

"This is our model moving forward"

Although the Washington Metro LED is just one project, Philips does not expect it to be a niche offering, according to William McShane, Director of sales for the sustainable lighting division at Philips Electronics North America. "This is our model moving forward".[32] McShane said Philips was already in talks with other commercial entities and municipalities for similar style contracts. "This is driven by the market," he said. "They want LEDs and controls, but they don't have the knowledge or the in-house resources" to develop the end-to-end system and get financing. Philips wants to be the one-stop source that offers financing, technology, installation, and maintenance.

Philips is also extending its energy services to other countries. For instance, Philips China recently announced the establishment of Philips Energy Service Company in Wuhan. The company hopes to advance the promotion of high efficiency lighting solution through the co-operative resource business model. Four leading businesses from the Wuhan local area have signed an Energy Performance Contract with Philips.[33]

Managing Customer Relationship over Time

"The product is no longer the end", Clive Grinyer, head of customer experience with Cisco said. "The role of the product designer is increasingly one of interpreting cultural behavior, of translating customers' emotional requirements".[34] The third dimension of the construction of the offering as suggested here focuses on the management of the relationship with any individual client or customer over time. The idea is to address the

[31] http://www.greenbiz.com/blog/2014/01/27/philips-lighting-teams-mayors-power-led-use-cities.

[32] http://www.greentechmedia.com/articles/read/washington-metro-will-install-leds-at-zero-cost.

[33] http://www.ledinside.com/news/2013/9/philips_china_establishes_energy_service_company_in_wuhan.

[34] C Grinyer (2011). New challenges for product design: The product of design. *DMI Review* 22(4), pp. 15–21.

dynamics between the physical plane and customer service through a system of engaging relationship management that is attached to the delivery of the expected value associated with the so called "product". Companies are realizing that customer service is a lever for differentiating a product, and can ensure the continuity of a relationship between user and manufacturer throughout the period from equipment purchase, or lease or other forms of distribution, to its renewal and much beyond.

As part of the general context of changing consumer expectations, service support operations provide an avenue for establishing an ongoing relationship through which the two parties combine their abilities in order to jointly ensure that a product fully attains its intended usefulness and much more. The customer, ideally, fulfills his or her aspirations of getting extra value, such as additional efficiency, peace of mind, additional acquired knowledge, inspiring thoughts, and the like, from the "product". He or she consumes an articulated set of "service" over time and pays the "service-manufacturing" supplier in a variety of ways. This relationship helps maintain feedback from the field to the design and production system about the evolution of customers' needs, and the reliability of the product and services. To do this, effective manufacturing and retailing firms tend to reinforce their support networks to create, when possible, the conditions for individualized customer follow-up.

In a traditional firm, the lack of inter-functional co-ordination leads to the two separate paths resulting in separate designs, with the subsequent iterative, inefficient, and time consuming realignment of the physical dimension and the service dimension into the definition of a consistent "package", combining tangible features and appropriate sets of services. The firm that recognizes the fundamentally cross-functional nature of product/service design, and can reduce the usual organizational conflicts, is more likely to take the direct (although rarely linear) path from the launch of the new product to the design of the coherent package submitted to the market. There is little new in the above discussion. However, it ignores the fact that product characteristics, as well as customer needs, may change over the course of the product's lifetime. Hence, the time axis may also change. Consider, now, the following activities:

- Managing the "physical" component of the "product" over time: The technological description of repair, maintenance, reconditioning, or re-manufacture of the product over time, and/or the specification of future upgrades or add-ons and/or decisions as to the compatibility of the product with other products, present and future.

- Managing the "service" component of the "product" over time: The anticipation of how the customer's use of the product evolves with time, including the customer's changing needs from the product in conjunction with other available products.

Again, aggravated by insufficient inter-functional communication or co-ordination, these activities may happen independently, leading to the problem of a lack of coherence of the end result. Furthermore to be effective, the process must match product properties over the course of its lifetime, with changing customer needs, in order to arrive at an ultimate definition of the company's "value proposition". This process will also lead to the planning of a variety of ancillary activities: Field service, parts inventories, etc. As we discussed above, the iterative "matching" that occurs as "physical component managed over time" and "service component managed over time" gradually coalescing into the "ultimate value proposition", is a spectrum of the inter-functional conflicts that exist in most firms. Firms that recognize the need to bring a multi-functional team to the design process are able to move along a direct path to the end result. The very compartmentalization of activities along the three suggested axes shows how ineffective (and how potentially destructive) is the functional compartmentalization typical in so many corporations: R&D deals only with its narrowly defined tasks, as does Marketing, as does Field Service, and so forth. It then becomes difficult, if not impossible, to develop products that will satisfy customers over the products' lifetime.

From such a perspective one might propose a categorization of the major strategic decisions that must be made in defining a Service Mix policy. This is only a very succinct guide. These steps are not meant to be sequential. In a very real sense, these decisions must be made simultaneously; since their interactions may be of great impact.

First Step: Design the physical product so that it is adapted to service objectives and constraints:

- Determine the technical parameters of the tangible product while taking into account not only its performance when new, but also its serviceability and maintainability.
- Manufacturing decisions in relation with reliability and durability.
- Make decisions about modular design and compatibility with other equipment.
- Make decisions about the integration into the product of internal diagnosis and/or remote maintenance systems.

- Determine whether or not the item has to be designed for repair or remanufacture either once or several times during its life.

Second step: Design the service package, and determine the range and performance level of services to go along with the tangible product:

- Determine the range of services to be offered, e.g., the balance of Service Mix components.
- Segment the Service Offer after identifying the significant segmentation criteria and analyzing the relative profitability of different segments of the Service Mix activity.
- Propose support for competitors' equipment after analyzing the opportunity to increase market share by providing this kind of support.
- Determine the level of performance for each component of the Service Mix offer.
- Allocate the appropriate level of investment into each service activity.

Third step: Manage the product/service throughout its lifetime, taking into account some of the following economic and organizational issues:

- Profitability objectives and the timing of payments and expenditures.
- Strategic alliances with other firms for the delivery of some of the services.
- The appropriate organizational structure for the Service Function.
- Human resource management within the Service Function.

REINVENTING REMANUFACTURING

Along with improvements in the standard of living, energy consumption and the consumption of other non-renewable material resources have grown to an unsustainable level. Products are consumed and then disposed of after use. Remanufacturing presents a huge environmental and financial opportunity for sustainable prosperity. Simply stated, remanufacturing is the process of disassembly of products wherein parts are cleaned, inspected, repaired or replaced and then reassembled to 'like-new' condition. The part or product is then returned to service in a system. Product or component remanufacturing, reuse, or recycling provide for material recirculation loops within the product system. These material recirculation loops reduce the requirements of raw material input and the overall energy input, and also reduce the product system waste. Remanufacturing typically uses 85% less energy than manufacturing.[35] The Fraunhofer Institute estimated that raw materials saved by remanufacturing worldwide each year could fill 155,000 railroad cars forming a train 1,100 miles long.[36]

Besides the environmental opportunity, another important advantage of manufacturing is its economical efficiency. In fact, in the US, the remanufacturing industry was primarily developed not due to environmental concerns or government regulations, but because it is economically efficient.[37] Compared with producing a new product, remanufacture results in substantially reduced energy and virgin raw material use, and production cost reductions. Hence, if product design permits, and a sufficient process for return of used products (e.g., reverse logistics chain) is ready, there is a lot of potential for a strong business model in product remanufacturing. Lavery and Pennell conservatively estimated a net reduction of approximately 34% in input costs. With reduced input costs and increased labor spent, there can still be up to a 50% increase in gross profit.[38]

Design for Remanufacture

The largest gain in enhancing reusability and the ability of remanufacture can be made in product design. Design directly impacts on a company's ability to monitor, disassemble, recover and reassemble products, and has been viewed as the most critical factor for enabling remanufacturing. Products need to be designed with the circular economy in mind. "Design for Remanufacture" is a combination of design processes whereby the constraining issues are reduced and items are designed to facilitate remanufacture. Design for Remanufacture can therefore play a role in optimizing the process of remanufacture.[39]

Design for Remanufacture is guided by an assessment of product or component value over time. It is important that the life-cycle management issues are carefully considered during the product development. The product should be designed in such a way as to support remanufacturing,

[35] R Steinhilper (2006). *Remanufacturing: The Ultimate Form of Recycling.* Stuttgart, Germany: Fraunhofer IRB Verlag, p. 6.
[36] http://www.fraunhofer.de/en.html.
[37] http://www.mech.kuleuven.be/lce2006/key4.pdf.
[38] Lavery G., Pennel G., Brown S. and Evans S. (2013). The Next Manufacturing Revolution, http://www.nextmanufacturingrevolution.org.
[39] N Nasr and M Thurston (2006). *Remanufacturing: A Key Enabler to Sustainable Product Systems.* Rochester, New York: Rochester Institute of Technology.

reuse, and/or recycling. The detailed product design issues include:

- Design for disassembly (and separation).
- Design for multiple life-cycles (cleaning, product reliability, durability, and remediation).
- Modular design: Functional clusters of components with similar technical (durability) and market life (technology change rate).
- Design for evaluation (embedded condition or usage monitoring).

In the traditional case, manufacturers mainly focused on making products. After the product is sold to the consumer and beyond the warranty period, the product manufacturers no longer have a role in the product life cycle. But now in the service economy the case is different. The product manufacturer plays a leading role in the entire product life cycle, including design, manufacture, service and support, retirement, and disposal). In such case, efficient material resources use and reuse, and the value associated with the reduced energy are taken into account. Design for Remanufacturing can therefore generate considerable economic returns to the product manufacturer, and also produce a more sustainable product.

Organizational Zoom . . .

Balzers Tools Bits: Product-Service Remanufacturers[40]

A tool bit manufacturer, Balzers, has moved into a new territory in the very traditional tooling industry. Once seen as ordinary consumables, tool bits are now considered as highly remanufacturable components. Thanks to remediation technologies, it becomes possible now to recoat and grind the tool bits to 'as-new' condition. These are coupled with convenient dispenser systems, such as supply and return vending machines, or intelligent hole-count machine tools, at client companies such as Ford. Benefits include improved tool productivity management, overall reduced cost, and also conversion of capital cost into variable cost, thus improving the cost allocation.

Some manufacturers have recognized that through leasing the product to the end customer, the selling of a service can incentivize better design. In 1962, Rolls Royce developed a "Power by the Hour" model to sell its aircraft engines. By offering a complete replacement service on a per hour

[40]http://www.remanufacturing.org.uk/pdf/reman_primer.pdf.

basis of use, mutual benefits are received. Customers are guaranteed a fixed engine maintenance cost over an extended period of time, while on the other end, Rolls Royce keeps the design and associated value of remanufacture in-house.

Design for Remanufacture approach prevails and thrives mostly in sectors of very high value or technological content such as aerospace, military, and power turbines. It is also valid in sectors that have embraced the concept of Product Service Systems, since in these instances there are shared motives for product longevity, durability, and performance.[41] Yet, Design for Remanufacture is not a "one size fits all" approach, and in some cases will not be the best option, either economically or environmentally. For example, remanufacturing is least prevalent in areas producing fast-moving consumer goods where fashion or status-related purchasing decisions are highly relevant and technology moves quickly, and it may be better to encourage other strategies such as recycling or reuse in these circumstances.

Organizational Zoom . . .

Caterpillar: "As good as new, as strong as ever"

Caterpillar is an American corporation, which designs, manufactures, markets and sells machinery and engines. In 2013, with sales and revenues of $55.656 billion, Caterpillar became the world's leading manufacturer of construction and mining equipment, diesel and natural gas engines, industrial gas turbines, and diesel-electric locomotives.[42] The company is also a leading service provider through Caterpillar Financial Devices, Caterpillar Remanufacturing Services, Caterpillar Logistics Services, and Progress Rail Services. It supplies and remanufactures military equipment. All types of military tanks as well as Land Rover Snatch models are remanufactured by Caterpillar, which in turn enhances the military's green credentials. For a typical Caterpillar product, 10% of the product is remanufactured with 90% remaining unchanged.[43]

Caterpillar's remanufacturing business began in 1972, and has expanded exponentially, following demands from its customers in the US for high

[41] Parker, D. Remanufacturing in the UK: a significant contribution to sustainable development? RRF, North Yorkshire, 2004. http://www.remanufacturing.org.uk/free-reuse-publication.lasso?report=48

[42] http://www.cat.com/en_US/news/engine-press-releases/caterpillar-launchescatreman propulsionenginesforyachts.html.

[43] Report: Remanufacturing Towards a Resource Efficient Economy. http://www.policy connect.org.uk/apsg/report-remanufacturing-towards-resource-efficient-economy.

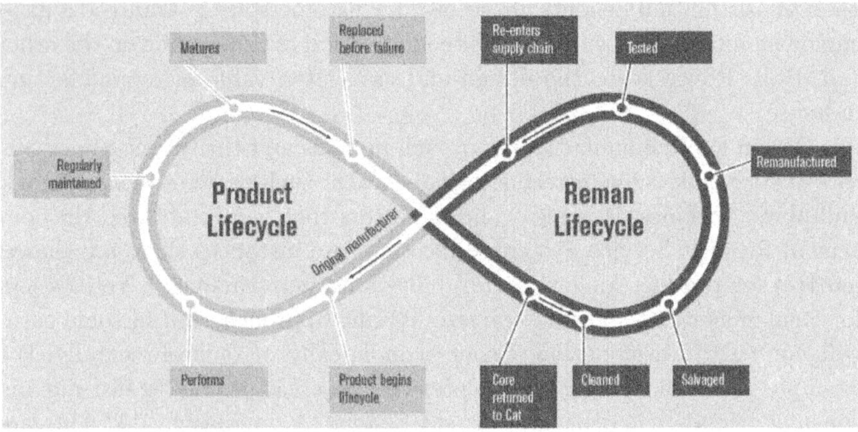

Figure 4.3: Product and Remanufacture Lifecycles for Caterpillar Products.[44]

quality, low cost replacement engines for their on-road truck fleets. Caterpillar realized that by keeping products in-house and remanufacturing them, they would retain ownership of their products and the associated value.

Through its highly controlled "Cat Reman" brand, the company has increased profit margins while still producing components of the highest quality, by replacing products before they break and rebuilding them with a mixture of new and used parts, as illustrated in Figure 4.3.

35% of Caterpillar's costs lay in overheads and 65% in materials costs. Salvaging materials therefore provides it a competitive advantage, compared with companies that solely focus on driving down overheads.[45]

To hold on to its core, Caterpillar created a returns incentive scheme. The cost of a remanufactured part, or "core", is equivalent to the cost of a new part unless an old part is returned, in which case it is cheaper. Essentially, the customer pays a "core deposit" which is refunded upon return of the core (provided it is complete and has no extreme damage). To Caterpillar, this economic incentive helps the company to receive a high percentage of "core" back for remanufacturing. In addition, the scheme reduces the material costs and allows Caterpillar to retain control of materials at the end of the traditional product life. The scheme also ensures fair pricing: Customers only have to pay the remanufactured price after the "core deposit" is returned.

---

[44] *ibid.*
[45] *ibid.*

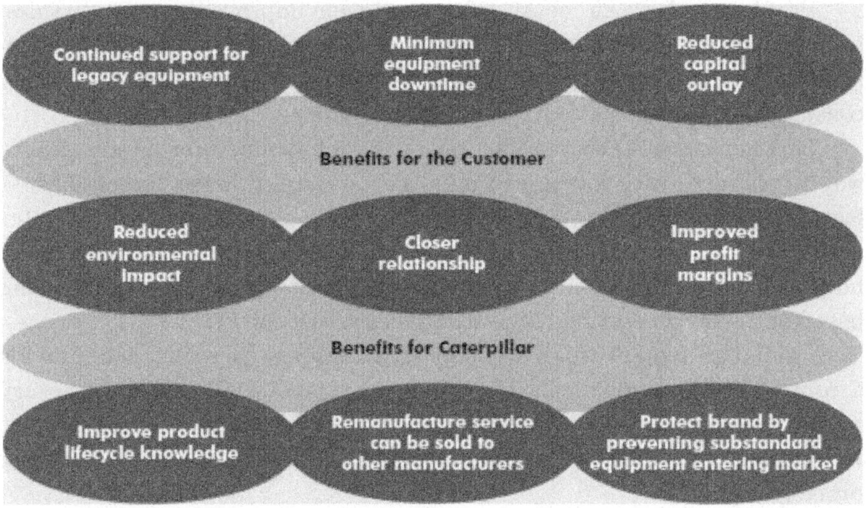

Figure 4.4: Caterpillar Remanufacturing Services: Business Benefits.[46]

Caterpillar's remanufacturing services also have extensive environmental benefits. The company has calculated that remanufacturing a cylinder head leads to an 86% safety advantage, 61% less greenhouse gases, a 93% reduction in water-use, an 86% reduction in energy-use, a 99% reduction in waste sent to landfill, and a 99% reduction in material-use compared to making a new product.[47] Figure 4.4 shows how remanufacture benefits both Caterpillar and its customers.

The motto for its remanufacturing division says, "as good as new, as strong as ever". Indeed, every remanufactured product that leaves the factory has been through rigorous quality test procedures, often having been passed along the same production line as a new product. This is backed up by a full warranty, the same as is issued with a new product.

Remanufacturing as a Service

To derive maximum benefit, remanufacturing depends on the customer recognizing the value of his end-of-life asset, and the manufacturer creating products that have the required durability in critical components. Both these motives can be enhanced if the goods can be seen less as products and more as a service.

[46] http://www.remanufacturing.org.uk/pdf/reman_primer.pdf.
[47] EEF: The Manufacturers' Organization.

Profit margins can be sustained or even improved over new sales, particularly if performance upgrades can be included in the remanufacturing activity. Customers can then stay up to date with respect to energy consumption, emissions controls or safety systems.

In other cases "servitization" offers the remanufacturer an alternative way to compete with low-priced imports. To date, this has largely been a feature of business to business services, where in general, there is a greater awareness of financial implications of purchasing decisions. Even here, however, there are plenty of opportunities for more informed purchasing based on life-cycle costing rather than "least cost, now".

From the manufacturer's perspective, remanufacturing can provide a route to address demands for producer's responsibility. The necessary recovery systems are accountable and demonstrably more resource efficient when integrated with the management accounting of the manufacturing processes.

AN INTEGRATIVE APPROACH TO NEW PRODUCT DEVELOPMENT

Under the suggested new paradigm of product development process, service strategies should be formulated in parallel with design and development of the tangible product. Anticipating the "service needs" associated with the product lifetime leads to consideration of how the physical characteristics and parameters of the product match service objectives and constraints, just as designing the item itself leads to expressing specific needs in terms of technical considerations. Consequently, managing the product during its lifetime, in terms of service needs, is impossible to untangle from managing the lifetime of the tangible product. Over time, managing the interactions of the product evolution with the evolution of the service provision requires a specific approach of customer relationship. An anticipated management system has to be implemented to effectively monitor the relationship over time.

If the different sets of design activities are performed simultaneously but separately, there will in general, be some lack of coherence between the "endpoints" of these endeavors. Instead, there must be some "redesign" that matches tangible with intangible, leading to the point of convergence. Arriving at this third point also involves the myriad of activities that ensures the product can be manufactured to specification, can be delivered to the customer's premises and/or installed so as to be ready for use, and that the customer will be trained in its use. Often, the practical means used to translate service performance objectives into concrete terms, lead firms to simultaneously pull in opposite directions; more integration of the after-sales function with other functions of the firm; and simultaneously more separation, and often, the transformation of the after-sales function into a relatively independent profit center, or even into an outside subcontracted service. In some industries, allowing after-sales activities to not only be managed, but also planned, outside the firm is tantamount to giving away a source of competitive advantage.

In a depressed or highly competitive market, the outcome of an inadequate balance between service function integration and separation brings about conflicting objectives between the different parties involved in the "Service Chain": Marketing, Sales, R&D, Production, Supply, and After-Sales, etc. This may lead to the emergence of internal tensions that may progressively intensify. This would naturally reduce the effectiveness, and the profitability, of direct and indirect service support activities. Furthermore, the allocation of resources for, and within, the service function becomes more and more difficult.

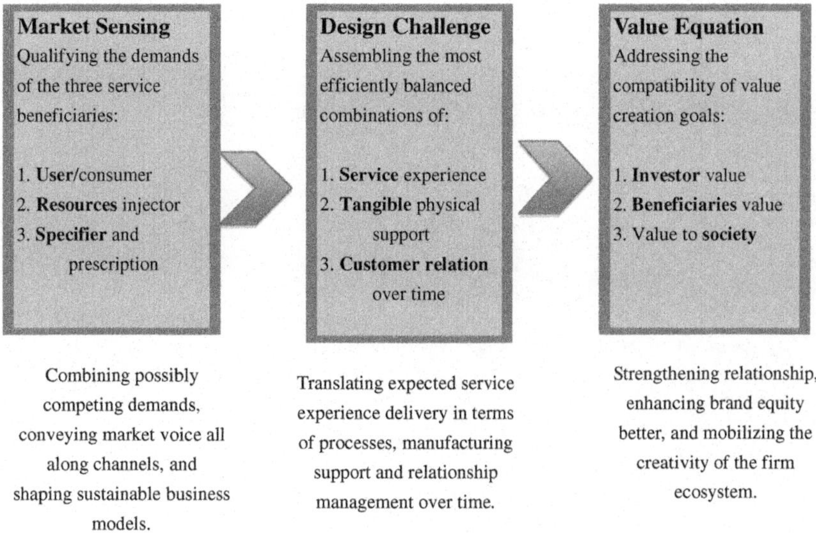

Figure 4.5: Combining the Three dimensions of Product Development in the New Paradigm of Innovation in the Service Economy.[48]

[48]H Mathe (2010) How Service Innovation Shapes the New World. In Moon-Joong Tcha (Ed.). *The Service Sector Advancement: Issues and Implications or the Korean Economy*. Seoul, South Korea: Korea Development Institute, and the Ministry of Finance and Strategy.

Figure 4.5 illustrates the three key stages of "product" definition as in the new paradigm of innovating in the service economy. As explained in Chapter 3, the first step focuses on the qualification of the demand for the beneficiaries of the value being delivered: the user or consumer; the resources injector or "paying" body; and the specifier who possibly "prescribes" the use of the "product". As described in Chapter 4, the second step aims at assembling the most efficiently balanced combination of service experiences; physical supports; and customer relation management processes. The third step, as developed in the coming chapters, addressed the shaping of the value equation. Value has to be created for the investor, as well as for the beneficiaries as identified in step one, and also to satisfy societal needs and requirements.

Chapter 5

THE IMAGINATION ECOSYSTEM, BEYOND OPEN INNOVATION

Open innovation is a concept usually attributed to Henry Chesbrough, a professor at the University of California in Berkeley who established the Center for Open Innovation. Chesbrough defines Open innovation as "*a paradigm that assumes that firms can and should use external ideas as well as internal ideas, and internal and external paths to market, as the firms look to advance their technologies*".[1] Although Henry Chesbrough was the first to coin the term in his book "Open Innovation: The new imperative for creating and profiting from technology" in 2003, the concept of open innovation existed long before he labeled and drew worldwide attention to the process.

Napoleon and "Open Innovation"

Open innovation is increasingly employed by today's companies. When the controversial French leader, Napoleon Bonaparte, employed the same technique in the late 18th century it was considered a revolutionary concept at that time. Realizing that almost as many of his soldiers died from scurvy and malnutrition at the hands of the enemy, Napoleon offered a 12,000 franc prize for anyone who could come up with a method of preserving food to feed his men. Indeed, Napoleon is famous for

[1]HW Chesbrough (2003) *Open Innovation: The New Imperative for Creating and Profiting from Technology*. Boston, MA: Harvard Business School Press, p. 24.

saying *"An army marches on its stomach"* and devised the creative method of creating an incentive to source this critical information. The incentive drew the interest of many but it was Nicolas Appert, a French confectioner and brewer, who scooped the prize offered by Napoleon. Appert invented a new method of food preservation which consisted of storing food in glass jars, sealing them with corks, and subsequently heating them in water.[2] What Appert gained in money, Bonaparte gained in the development of a technique that otherwise may have taken years or even decades to be invented. This is indicative of the potential value of open innovation which accelerates the speed of development of innovations.

Essentially, open innovation entails a firm to open its communication channels to learn from players or users outside the firm, in order to enrich the innovation capacity of the firm, and is a dynamic collaborative concept that can be applied to product development, distribution or market launch. As Rae (2008) confirms, "*open innovation is about connecting with others to find new ideas and, often, to co-develop and co-market them*". Today, innovation ranges from pan-industry innovation networks that share in the risks and rewards of their findings to straightforward co-development projects between strategic players; to buying or licensing inventions and collaborating with companies and customers. In short, innovation as a concept is becoming progressively more open and companies are discovering new ways to enrich their new product development by listening, learning or collaborating with the outside world.

To understand more deeply the concept of value exchange that underpins open innovation, David Ricardo, the early 19th-century economist analytically clarified the classic trading system of exchange. What one company lacks in expertise can be sourced and exchanged for a price from another company in order to prevent the need to invest in expertise that is entirely new to the company. Paul R. Lawrence, the Wallace Brett Donham Professor of Organizational Behavior at Harvard Business School illustrates the fundamentals of this dynamics: "*Imagine that tribe A is good at both hunting and fishing, but more efficient at hunting. Tribe B is not as efficient as tribe A at either hunting or fishing, but is more efficient at fishing than at hunting. Ricardo pointed out that both tribes would eat somewhat better if tribe A sold game to tribe B in exchange*

[2]L Day and I McNeil (1996) *Biographical Dictionary of the History of Technology*. London: Routledge, Taylor & Francis Group.

for fish. Both parties would win; most notably, the poorer one".[3] Such is the case with open innovation whereby knowledge that cannot be created within the firm is sought from a source of expertise outside of the firm.[4] The convergence of complementary or exchangeable expertise by several parties, known technically as *"distributed co-creation"*, enables the exchange of useful information that helps both parties to advance their expertise.[5]

Yet, why are many companies so reluctant to embark upon a strategy of open innovation? It may help to understand the degree of openness or the type of open innovation that is appropriate to the firm. Companies such as Procter & Gamble (P&G) have achieved great success at the open innovation model "Connect & Develop"; but the process of innovation should be matched to suit the type, size, and culture of the company, and should ensure that power over the firm's intellectual property is not relinquished to an outsider.

> *"In an era when great ideas can sprout from any corner of the world and IT has dramatically reduced the cost of accessing them, it's now conventional wisdom that virtually no company should innovate on its own".*
>
> — Gary P. Pisano and Roberto Verganti[6]

Gary P. Pisano and Roberto Verganti recommended companies to consider four basic collaboration approaches. Their proposed approaches differ along two dimensions: Openness (is it a completely open system or restricted to selected players?) and hierarchy (who calls the shots — one lead participant or is it a consensual process involving all players?).

According to these two aspects, there are four basic modes of collaboration: A closed and hierarchical network (an elite circle), an open and hierarchical network (an innovation mall), an open and flat network (an innovation community), and a closed and flat network (a consortium). In the open, hierarchical approach, anyone can offer ideas but the company defines the problem and chooses the solution. In the open, flat approach, anyone can solicit and offer ideas, and no single participant has the authority to decide

[3] P.R. Lawrence (2010) The drive to acquire impact on globalization. *HBR Working Knowledge.* Aug 23 - hbswk.hbs.edu/item/6415.html
[4] *Ibid.*
[5] J Bughin, M Chui and B Johnson (2008) The next step in open innovation. *McKinsey Quarterly.* Number 4. pp. 113–122.
[6] GP Pisano and R Verganti (2008) Which kind of collaboration is right for you? *Harvard Business Review.* December. Vol. 86, Number 12.

what is or is not a valid innovation. In the closed, hierarchical approach, the company selects key participants and makes the decisions on which ideas get developed. In the closed, flat approach, a handpicked group is invited to offer ideas, but participants share information and intellectual property and make critical decisions together.

We share the view of Pisano and Verganti that to select the best option, a company needs to carefully evaluate the advantages and disadvantages of different collaboration approaches depending on the strategy and capabilities of the company. Yet, taking the standpoint of the focal company, we especially emphasize on companies' existing knowledge and assets. We developed a relatively simple framework for top management, such as Chief Innovation Officers, to make better decisions about their engagement in open innovation and types of collaboration their companies implement. It mainly concerns two questions: Does the company have the capability to identify the problem and decide which solution to adopt? And does the company have the knowledge and expertise needed to solve the problem in hand?

Companies differ in the extent to which the capabilities, that define the problem and evaluate the proposed solutions, are concentrated within the organization. In some cases, the authority to decide which problems are most important, what constitutes an acceptable solution, and which solutions should be adopted is completely vested in the focal company. In other cases, companies need to install the necessary capabilities with some external assistance to make the judgment. Examples include companies such as Zara and Lego which develops collaborative projects with users and customers in order to decide future products.

Companies also differ in the extent to which they possess the knowledge and assets that are critical to the sought-after innovation. Some companies have the ability to conduct R&D or create innovation internally, while some companies do not have the necessary breadth of perspective or capabilities. In the latter case, companies need to source external partners to cultivate the solution knowledge and development capabilities. With the advancement of technology, potential partners and ways of collaborations have both dramatically increased. InnoCentive, for instance, is one of the platforms that connect a large pool of individual Solvers with a number of Seekers organizations.

Based on the two dimensions, there are four basic modes of innovation pattern, as shown in Figure 5.1: Internal development, innovation tournament, innovation community, and market monitor. Each of these four modes will be illustrated with examples in the following sections.

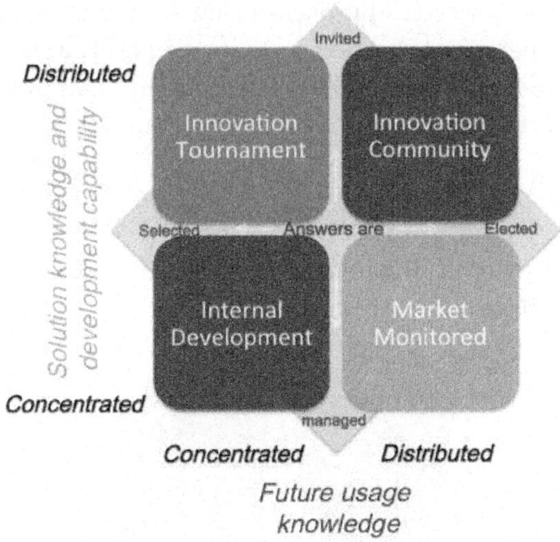

Figure 5.1: Collaborative Innovation Approaches.

STIMULATION FROM WITHIN

Internal stimulation means that the company develops the new product with its own capabilities. The company bears all of the costs associated with new product development and implementation while mobilizing the talents and infrastructures of its internal research units. This approach makes sense when a company trusts its ability to anticipate future use or consumption of a future product or service, and also possesses the capability to internally run the development phases associated with the projects. However, network value may still operate when companies implement collective abilities to connect with different parts of the world and with varied fields of expertise and technological excellence.

Internal Development

Innovation by relying on internal resources and decision-making processes is considered to be a relatively conservative approach. Nowadays, companies have created new ways to stimulate the potential of their internal resources. Some companies have established successful entrepreneurial programs that encourage employees to think up and develop new product ideas. For example, 3M's well-known 15% rule allows employees to spend 15% of their time in bootlegging, i.e., working on projects of personal interest whether or not those projects directly benefit the company. Companies sometimes look for creative innovation approaches that overcome barriers to the free flow of new product ideas. For example, firms like Eureka' Ranch, a well-known new product hatchery employ both method and madness in helping companies to start their product idea generation process.

Organizational Zoom . . .

To Boost Internal Innovation, Ericsson Thinks Inside the Boxes[7]

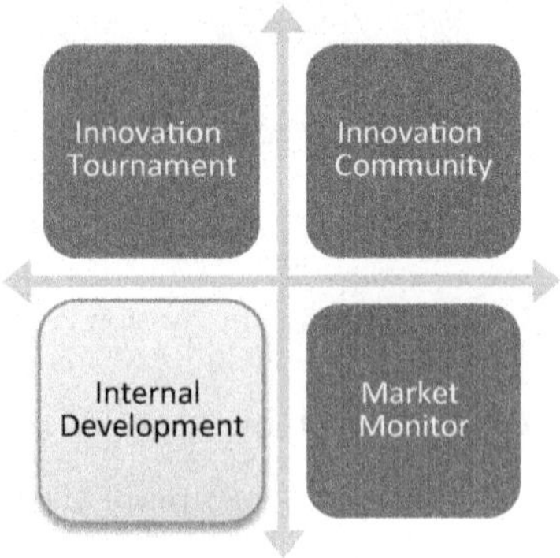

Ericsson, the Sweden-based communications technology company, has a long history of technology and business innovation in communications, leading the industry in areas such as digital switching, mobile systems, and management services. Going forward, innovation must be accelerated to meet increased competition and broadened to put a greater emphasis on areas such as services and business model innovation.

In the overall strategy, market and insight driven innovation — typically in close cooperation with users, customers and partners — is complementing traditionally strong internal research and technology-led innovation. The strategic emphasis on innovation is supported by the overall Ericsson mission statement: "Innovating to Empower People, Business and Society, the long term business strategy and clear internal messages from the senior team that all employees are expected to contribute to the innovation efforts".

The innovation strategy includes widening the front-end to source ideas broadly internally and externally, establishing a network of connected

[7]http://www.managementexchange.com/story/everyone-innovates-everyday-collaborative-idea-management-ericsson.

innovation funnels at different levels across the organization, including a corporate incubator addressing business opportunities beyond "business as usual" at Group level, as well as defined strategic growth areas, relevant performance indicators and an extended innovation toolbox.

In 2008, Ericsson started to implement a collaborative idea management tool called IdeaBoxes. The key innovation of the approach, as well as a major success factor, was the pull based and self-organizing nature of the idea management system.

With many users and a large number of ideas, complexity is increasing. Also, in large organizations such as Ericsson, there are many and diverse innovation needs distributed across the organization. This means that it is necessary to design a structured approach to channelize ideas from multiple sources to multiple destinations in the organization.

A compelling solution was to build a scalable and self-organizing system where innovation needs can be activated as they are identified and where ideas are automatically matched towards the existing innovation needs at any given time. In this way, collaborative idea management was implemented like a pull-based internal idea marketplace: An open network for the exchange of ideas built around a host of defined innovation needs, without any central control or steering.

The system is open to all employees within Ericsson globally and participation has grown gradually "bottom up", through viral marketing. It is up to each innovation manager to decide if they want to use the tool to help them gather and manage ideas relevant to their innovation efforts. Since the launch of IdeaBoxes, there has been a broad adoption throughout the organization across all business units, market regions and group functions, and in all areas including R&D, sales and marketing, and internal processes etc.

The use of the tool has been promoted by several focused idea generation campaigns, hosted by different internal units but open for all employees. In mid-2011, IdeaBoxes had grown to become the *de facto* idea management tool across Ericsson, replacing most local tools. After three years of use, the pull-based and self-organizing system has over 15,000 ideas. More than 300 innovation managers globally have opened idea boxes to attract ideas for their specific innovation efforts.

Internal Networks

Companies have realized that innovation should not be just R&D's responsibility; it should be a collaborative process involving customer, sales, finance, customer service, and operations, etc. Internally, the company can pick the brains of its executives, scientists, engineers, manufacturing staff, and sales people. To foster the very culture of collaboration across corporate functions and departments, some companies are actively developing internal innovation networks and institutional platforms. On the one hand, such internal innovation networks bring together corporate units (R&D, business development, and marketing, etc.) to communicate on strategic issues. They serve to select and define R&D problems that are most important. On the other hand, internal networks provide intellectual and operational support for developing and implementing new product ideas. In particular, internal networks have the advantage of finding solutions to problems that individual units have failed to solve by themselves. Such an internal brainstorming may turn to be especially productive in multinational corporations with numerous units spread across countries and continents. People in such distant units, often created as a result of mergers and acquisitions, rarely communicate with each other and almost never meet face to face. Yet, people in one unit may possess a specific knowledge that is desperately needed and can be immediately used, in another. Connecting these people through internal innovation networks may therefore result in significant savings of time and money for internal R&D.[8] In addition, internal innovation networks can help to identify the emerging thought leaders, who especially if in junior positions would otherwise remain unnoticed in "remote" labs and cubes.

Companies may choose to build internal innovation networks in different forms to suit their specific needs. One of the best-known examples of internal innovation programs is Qualcomm's Forward Looking User Experience (FLUX), an employee-run, cross-departmental innovation club that brainstorms novel solutions to the company's existing problems and also identifies new problems. Launched by only eight people, this program now includes 28,000 Qualcomm's employees, yet exists on very limited budget and employs no single full-time equivalent (FTE).[9]

Other companies choose using various innovation management software and web-based platforms. Coming in different shapes and shades, these tools seek to capitalize on modern trends in social media. A case in point is Ericsson's IdeaBoxes, an online idea sharing platform that is used for all types of ideas, from game-changing to incremental. Employees can post

---

[8] http://www.innovationexcellence.com/blog/2013/07/13/internal-innovation-networks-foundation-for-success/.

[9] http://www.innovationexcellence.com/blog/2013/07/13/internal-innovation-networks-foundation-for-success/.

their ideas to an online forum to solicit peer feedback and votes of approval. The system has helped generate over 30,000 ideas, and one out of every 30 ideas generally proves good enough to get implemented.[10]

SOURCING DISRUPTIONS

The "open" approach to innovation has risen extensively in usage in recent years with increasing numbers of companies becoming aware of the value that the wider market has to offer in terms of fueling innovation efforts and sourcing ideas. Some companies have a clear understanding of the future trend of a particular product or service development (markets and user needs), but they may not possess the necessary solution knowledge and development capability within their organizations. For such companies, the closed approach that regards the expertise and knowledge within the company as the privileged source of innovation would be not sufficient. Instead, they may turn to external market place, usually virtual, which can potentially attract a large group of talents from outside the company, and consequently, obtain needed services, ideas or content. First, the Seeker organization may post a critical question in the open market place. Organizations do not need to identify either the best knowledge domains or the most appropriate experts in those domains. Then it is just like holding an innovation tournament, i.e., Solvers, wherever their location, compete by proposing solutions, and the Seeker organization chooses the solutions it likes best and rewards the Solver.

Innovation Tournaments

With open participation, companies take a function traditionally performed by employees or suppliers and outsource it to an undefined (and usually large) network of people in the form of an open call. Technological advancements in areas from product design tools to digital cameras are breaking down the cost barriers that once separated laypersons from experts. Nowadays, as smart companies in industries discover ways to tap the talent of the crowd, hobbyists, part-timers, and dabblers suddenly have a market for their contributions.

Compared with a traditional workforce, the advantages of the "innovation tournament" model include flexibility, scalability, access to a broad

---

[10]http://www.fastcompany.com/3004745/creative-conversations/boost-internal-innovation-ericsson-thinks-inside-boxes.

range of skills and experiences, with costs which are substantially less than those associated with more traditional forms of employment. It appeals to companies considering the low labor costs, overheads, and administration, coupled with the lack of employment regulations and employer security.

Organizational Zoom ...

P&G: Connect & Develop

"Connect + Develop Open Innovation has helped deliver some of P&G's leading innovations, and is critical in helping us deliver on our renewed growth strategy moving forward. Connect + Develop has proven to be a winning strategy, with still more room for growth".

— Bruce Brown
Chief Technology Officer

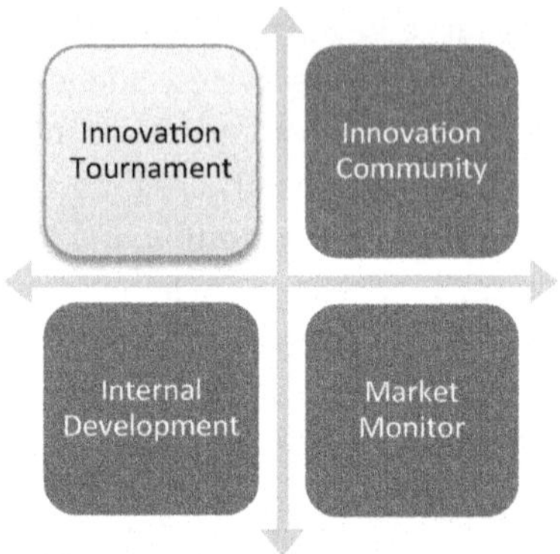

The Connect & Develop strategy by P&G is one of the most commented examples of a company soliciting the ingenuity and creativity of the world to devise a solution to a problem. The company faced the R&D challenge of finding a way to print directly onto Pringle crisps. As the expertise did not exist internally, this effort was incentivized with a financial reward and today boasts more than a 1,000 active agreements with innovation partners.[11]

P&G launched its Connect & Develop program in 2001. At that point, less than 10% of their new initiatives involved external innovation

[11]P&G Connect & Develop website: https://secure3.verticali.net/pg-connection-portal/ctx/noauth/PortalHome.do.

partnerships. By 2008, more than 50% of P&G products have benefited from some form of external collaboration. The Company even wants to take it further. The Connect & Develop programs aim to deliver $3 billion toward the company's annual sales growth by 2015.[12]

Since the launch of the programs, P&G has developed a broad spectrum of global partnerships, ranging from inventers (e.g., Crest Spinbrush), chemical suppliers (e.g., the pentapeptide at the heart of Olay Regenerist), University spinouts (e.g., the Bifantis® probiotic used in Align), competitors (e.g., Glad Press n' Seal developed with Clorox) to forming new business with other major organizations (e.g., the joint venture with Inverness Medical that became Swiss Precision Diagnostics).[13]

Some facts about P&G Connect & Develop[14]

P&G Connect & Develop partnerships

- More than 1,000 active agreements with innovation partners.
- Dedicated staff in over 10 countries worldwide seeking partnerships with universities/institutes, sole inventors, emerging companies, SMEs, MNCs, and competitors.

P&G Connect & Develop products

- Over 50% of P&G product initiatives involve significant external collaboration.
- Febreze- P&G's latest billion dollar brand relied on Connect & Develop for much of its success.
- Olay® Regenerist, Oral-B® Pulsonic® Toothbrush, and Pringles® all bring P&G Connect & Develop to life.

P&G Connect & Develop goal

- By 2015, external collaboration programs to be responsible for more than $3 billion per annum in new businesses.

To further speeden up and simplify external innovation connections, P&G's Connect & Develop program recently launched a new website (pgconnectdevelop.com). This website links innovators directly to top Company needs, which are categorized by business area and searchable

[12]http://www.pg.com/en_UK/news-views/Inside_PG-Quarterly_Newsletter/issue2/innovation3.html.
[13]*ibid.*
[14]*ibid.*

by topic. On the back-end, the site links P&G business leaders directly to external innovation submissions, letting them search by topic or work area, share internally, and track progress. The website has served as P&G's "open front door to the world", allowing any innovator anywhere to share their innovations with the Company.[15]

Crowdsourcing Platforms

Instead of creating their own open innovation programs, such as P&G's Connect & Develop, companies can also utilize the crowdsourcing platforms, including examples from InnoCentive.com for scientific problems, to OpenIdeo for social issues and design related questions. Organizations are realizing now that crowds are actually energized by intrinsic motivations such as the desire to learn or to burnish his or her reputation in a peer community.[16] In addition, although crowdsourcing is not always free, it costs much less than paying traditional employees. Also, Web 2.0 makes it fairly easy to engage these potential intellectual resources. By using the Internet, companies are just a mouse click away from an eclectic, university educated population ready to invest in intellectually stimulating projects for little or no remuneration.

Indeed, crowdsourcing occurs when a company uses a digital platform to leverage external resources for idea generation and exchange. Crowdsourcing and open innovation websites allow people from all cultural, economic, and educational backgrounds to collaborate on groundbreaking technology. This extends traditional outsourcing as companies use distributed labor networks to source external contributors, harnessing expertise on an unprecedented scale. An organization using crowdsourcing will set a challenge to the public and ask for solutions, opinions, and suggestions. It is an open call to the public whereby the company solicits solutions from the crowd. The current sophisticated online crowdsourcing platforms are making it even simpler to manage and support distributed workers.

Kickstarter is one of the most famous examples of an online crowdsourcing platform. It has grown to become the world's largest funding platform for creative projects. Every week, tens of thousands of people pledge millions of dollars to projects from the fields of arts, music, film,

[15] http://news.pg.com/press-release/pg-corporate-announcements/pg-connectdevelop-launches-new-open-innovation-website.

[16] KJ Boudreau and KR Lakhani (2013) Using the crowd as an innovation partner. *Harvard Business Review*.

design, fashion, games, technology, food, publishing, and other creative areas. Since its launch in 2009, over five million people have pledged in excess of $900 million to fund projects by creators who always maintain full ownership and complete creative control of their work.

Beyond the domain of raising cash for creative projects such as iPod dock and Bluetooth-enabled watches, a lot of other online crowdsourcing platforms are arising in recent years to solve diverse societal and technological problems. Kaggle aims at making data science a port, with a community of approximately 200,000 data scientist worldwide, from fields such as computer science, statistics, economics, and mathematics. It offers monetary rewards, jobs or consulting opportunities to statisticians and data miners who can make sense of problems with large data sets. Its partners include organizations such as NASA, Wikipedia, Deloitte, and Allstate for its competitions.[17] OpenIDEO hosts a small number of philanthropic challenges for social and environmental good, and it asks their online community to come up with new design solutions to societal problems like work health or affordable housing and education. The platform has seeded some great social impact, from helping a group of students at Stanford University increase the US National Bone Marrow registry by 115,000 people, to the launch of a health center in Columbia, to the design of an app rolled out globally by Amnesty International.[18]

Crowdsourcing is also an innovative way of developing new products. Threadless.com, a large online retailer of T-shirts, asks artists and designers from all over the world to submit designs for their products. About 1,000 designs are submitted online weekly and are put to a public vote.[19] Members of the "Threadless community" review and score the designs. Based on the average score and community feedback, about 10 designs are selected each week, printed on clothing and other products, and sold worldwide.

Organizational Zoom . . .

InnoCentive: Connecting Seekers with Solvers

One of the pioneers in the crowdsourcing arena is InnoCentive, a Waltham, Massachussetts-based Company, spun out from Eli Lilly in 2001.

[17]http://www.kaggle.com/about.
[18]http://oiengine.com/project/openideo-com/.
[19]https://www.threadless.com/make/submit/.

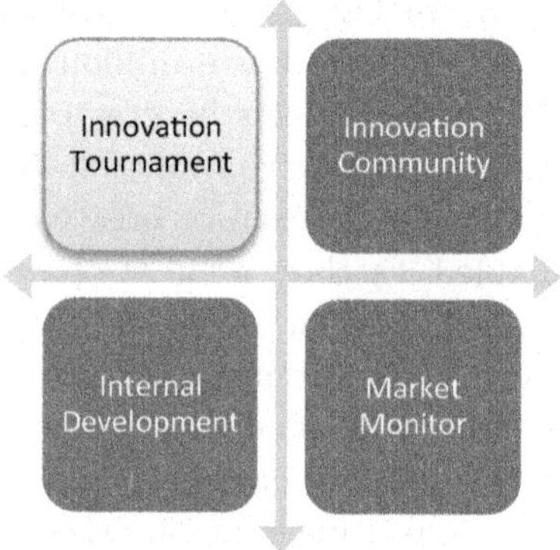

Essentially, InnoCentive was designed to connect a global network of individual Solvers with a number of Seekers organizations in short-term, standalone transactions. Seekers post a high-level problem statement, along with the associated financial award. The problems are generally smaller pieces of the seeker organizations' much larger R&D programs. Individual Solvers can view the challenges and volunteer to explore the problem and they would be provided additional information under a non-disclosure agreement to help them search for a solution. Solvers would then submit a proposed solution to be evaluated by the Seeker. If the Seeker is satisfied with the workability of the solution to a challenge, a pre-specified monetary award ranging from $5,000 to $1 million will be awarded to the Solver, in exchange for the acquisition of the intellectual property rights to the winning solution.

Bloomberg BusinessWeek reported that since its launch, "more than 170,000 participants from over 175 different countries have registered as Solvers. More than 800 problems have been posted, and almost 400 solutions have been found. This represents almost a 50% success rate on problems that had stumped internal R&D staffs. Almost $20 million in awards have been posted, while almost $4 million in awards have been paid out to successful Solvers. Given this success rate, InnoCentive has attracted a large and diverse set of organizations as Seekers, including Eli Lilly, P&G, Avery Dennison (AVY), Janssen, and the Rockefeller Foundation".[20]

InnoCentive's community of "Solvers" tackle problems ranging from engineering to life sciences, and business to IT, with sponsored prizes in the past from the likes of Merck, NASA and The Economist.[21]

[20]http://www.businessweek.com/stories/2009-04-08/the-next-wave-of-open-innovation businessweek-business-news-stock-market-and-financial-advice.
[21]http://www.theguardian.com/sustainable-business/crowdsourcing-open-innovation-change-world.

InnoCentive has also developed "InnoCentive at work", a cloud-based application that is designed to create "innovation communities" from groups of employees, partners, and customers.[22] This allows private crowdsourcing within corporations and agencies, giving them a cost effective way of collaboration and a much larger outreach.

InnoCentive has a community of competition approach whereby contributors compete to provide solutions and ideas in their respective fields. InnoCentive also prides itself on standing apart from other innovation providers in that customers only pay for results, not for work undertaken or failed ideas. InnoCentive calls this a "fundamental change in the economics of innovation".

Innovation Communities

If a company have neither the knowledge to decide the future usage of its products or services, nor the solution expertise and development capability, the "innovation community" approach would be suitable. Similar as the "Innovation Tournament" approach, an innovation community elicits the interest of "tinkerers", and retains the ability to tap the creativity and imagination of resources that lie outside the boundaries of the firm, and thus offers the potential to identify the critical problem and the capability to solve problems and find creative solutions to conundrums that could potentially not be solved alone. Yet, different from an "Innovation Tournament" where a focal company has the knowledge and capabilities needed to define the problem and evaluate proposed solutions, an innovation community would work well when no single company has the necessary scope of perspective or capabilities.

Innovation communication can be an open network where anybody can propose problems, offer solutions, and decide which solutions to use, or a consortium, where a closed group of members suggest questions, jointly decide how to conduct work, and possibly select the solutions. Open networks would be preferred when a problem can be partitioned into small, discrete pieces that individual players can work independently at a fairly low cost. Open-source software projects such as Linux, Apache, and Mozilla are examples of such open networks. They often develop very specific modules of code to address problems that users have encountered. In fact, users are usually the first to discover the problem, such as a bug in an existing

[22] http://www.evonomie.net/2014/04/30/marketing/is-crowdsourcing-the-future-of-business-innovation-5/.

piece of code or the need for a new feature. In such case, the users are best positioned to devise and test solutions, as they are closest to the problem.

Organizational Zoom . . .

Linux: The Free Open-Source System

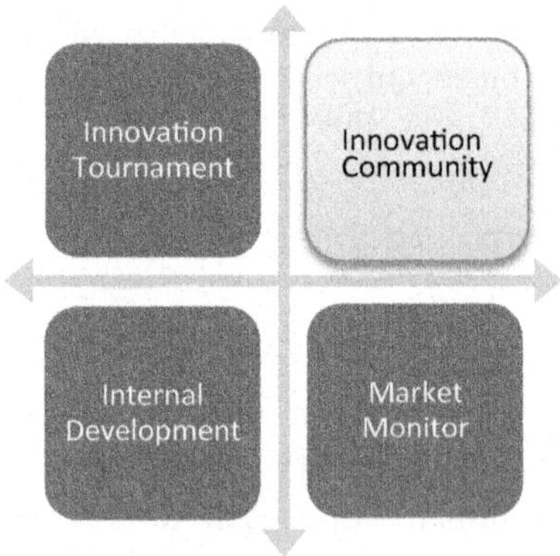

The Linux operating system, a free open-source system, owes its success to the collaborative development by a virtual community made-up of thousands of volunteer programmers and software developers motivated by a common goal. Instead of being motivated by money, Linux programmers claim that their goal is to make Linux an ever-bigger force in computing.[23] Linus Torvalds started the Linux phenomenon as a student at the University of Helsinki in 1991 with the help of fellow students and other iconoclasts who wrote code in their spare time. Today, the venture has become a cooperative project of massive proportions in which employees at some twenty or so companies, including IBM, Hewlett Packard, and Intel, along with thousands of individuals, work together to develop Linux software. The Linux community, according to Torvalds, is like a huge spider web, or multiple spider webs representing the dozens of related open-source projects that are developed collaboratively.[24]

Overcoming the Challenges of Collaboration

Collaboration, even when a mutual goal is at stake, is not always a smooth ride. What is it that makes the Linux collaborative ecosystem

[23]Hamm, S. (2005). Linux Inc., *BusinessWeek*, January 31. http://www.businessweek.com/stories/2005-01-30/linux-inc-dot.
[24]*ibid.*

work so effectively? It is an ecosystem driven by a shared motivation and guided by a culture of collaboration, meritocracy, and Darwinism. Torvalds said "*Any company or person is free to participate in Linux Inc., and those with the most to offer win recognition and prominent roles*". And as Torvalds states, "*I didn't design it this way. It happens because this is the way people work naturally*". Indeed, James F. Moore, a senior fellow at the Berkman Center for Internet & Society at Harvard Law School confirms, "Linux is the first natural business ecosystem".

Overcoming problems through consensus is central to the functioning and survival of such a shared system. One particular problem encountered at the embryonic state of the Linux, which came to be known as the "Twin Peaks truce", helped to cement the consensual relationship between programmers. In the early stages the founder, Linus Torvalds, was insisting on manually reviewing all additions to the software that were resulting in a "logjam" for the programmers and eventually incited them to rebel. The programmers threatened to develop a rival software package, something that likely would have crippled Linux, unless Torvalds changed his policies on reviewing their work. According to Larry McVoy, one of the programmers involved, "things were falling apart". But the crisis culminated during a meeting at McVoy's house on San Francisco's Twin Peaks when top Linux contributors urged Torvalds to change his mind. Finally, he relented. He agreed to delegate more and use a software program for automating the handling of code. The Twin Peaks crisis not only led to a more close-knit and democratic collaboration between the programmers, but it changed the way Linux was made and distributed. Overcoming a crisis such as this plays a key role in the survival of such a business ecosystem. A lawsuit filed by SCO Group Inc., a software company that claims IBM handed some of SCO's intellectual property to Linux, has rallied the collective spirit and a coordinated effort amongst Linux stakeholders like never before.

As an iconoclastic entity made up of independent contributors united by a common goal, Linux is rapidly becoming the strongest rival that Microsoft has ever faced. Now, Linux Inc. has emerged as a model for collaborating in a new way on software development, with important lessons for collaborative development and innovation throughout the business world.

Open networks reduce the risk of development and capitalize on diverse talents for a collective problem-solving endeavor. An example in point is Microsoft that is testing the merits of "friendly hacking" by unleashing the raw data to its *Kinect* game software. The company has left the door open for hackers to play around with the software and develop other uses for it. Hackers have created a website to record new uses for the device. The reverse engineering by hackers has led to the development of a robotic vacuum cleaner that is controlled with the wave of a hand, a model helicopter that senses walls and other objects in its path, and it has also found a use by surgeons. Kudo Tsunoda, Creative Director for Kinect says of the open innovation experience: "*When we launched Kinect, we knew we were only scratching the surface of what we could do and the possibilities of how it could grow over time were limitless. But we've seen so much innovation and creativity by people all around the world in building things with Kinect that even we couldn't imagine*".[25] Encouraging a developer community enables innovation to take place at a more rapid pace, and it also enlightens the companies as to the consumer value of such a technology. Microsoft has also developed Kinect Fun Labs — a community that allows innovations to be shared with others.[26]

A consortium approach would be efficient when the task involves developing radically novel product concepts or product architectures that cannot be separated into small, well-defined chunks. In such cases, employing a private group of participants provide an environment where members can closely interact and jointly decide how a specific problem is solved. For example, IBM has collaborated with a handful of selected partners to accelerate the development of semiconductor technologies and solve increasingly complex chip-design and manufacturing challenges. Although IBM is the leading member in this alliance, it is not the decision maker. Other partners including Free-scale, Infineon, Samsung, Siemens, and STMicroelectronics also have some say in how the technology is developed, as they possessed critical knowledge, skills, and assets that are essential for microprocessor design.

Open Innovation in Leading Organizations: A Consortium Approach

Dossia is a consortium of large employers including telecoms giant AT&T, the world's largest semiconductor chip maker Intel, the software and hardware manufacturer Pitney Bowes, and retail giant Wal-Mart

[25] C Nuttall (2011) Microsoft embraces hackers' use of Kinect. *The Financial Times*, June 6th.
[26] *ibid.*

which have come together to develop portable electronic medical records for their employees.[27]

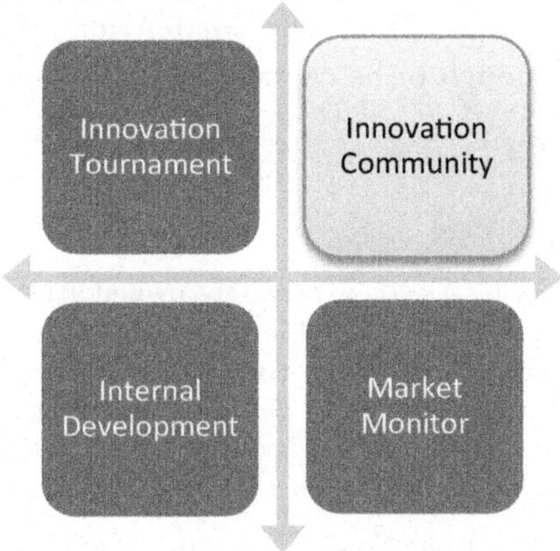

The Internet Home Alliance is a cross-industry network of leading companies such as home appliances manufacturer Whirlpool, IT giant Hewlett Packard, Cisco Systems, and Intel, engaging in collaborative research to increase use of Internet-based services at home.

Netflix, the Internet streaming company, has partnered with LG Electronics to create its own set-top box that will stream movies and other video content. The partnership is win–win: LG gets ready customers (who it expects will buy millions of its new boxes), while Netflix gets a new media platform that makes it more competitive.[28]

Bang & Olufsen has partnered with Samsung to create a co-branded mobile phone for the higher end of the market, and with BMW to develop its in-car sound system.

Collaborative design can extend beyond the metrics of the physical to the metaphysical value of brand. Mobile phone producers LG and Samsung have collaborated with two of Italy's luxury design houses, Prada and Giorgio Armani to lend a premium image to their handset devices. LG undertook an eight-month joint effort with designers at Prada and outpaced Apple's iPhone by about three months in the race to roll out a

[27] J Rae (2008) A ripe time for open innovation. *BusinessWeek*. http://www.businessweek.com / stories / 2008-03-19 / a-ripe-time-for-open-innovationbusinessweek-business-news-stock-market-and-financial-advice.
[28] *ibid.*

touch-screen model. The handset, called the Prada phone, is loaded with high-tech features hidden behind the black screen. Samsung's alliance with Italian designer Giorgio Armani also resulted in a new multimedia phone with a touch screen; the conveniently sized Armani phone of credit card proportions is light enough to be carried in a shirt pocket.[29]

Organizational Zoom ...

Innovation of Alessi

The design process in Alessi, a privately owned Italian design factory that specializes in accessories, took a different path in the 1980s, which has led the company to produce some of the most sought-after and outstanding houseware designs of the past few decades, including the first home espresso maker. Alessi broke with its traditional *"in-house-only"* design strategy and became one of the pioneers of open innovation.

Today the firm has relationships with about 200 external designers, from a multitude of sectors, many of whom are much better known for their expertise in fields such as architecture than in homeware design.

A teakettle by Frank Gehry and a vase by Zaha Hadid are just two products of those relationships. In addition to the typical process of issuing a design brief to designers, they regularly solicit the company whenever they develop an idea on their own accord. Alberto Alessi, head of his family's iconic design factory describe the process as thus: *"All of the 200 designers who work with us know they can call me and say, Alberto, I have a fantastic idea for you"*.

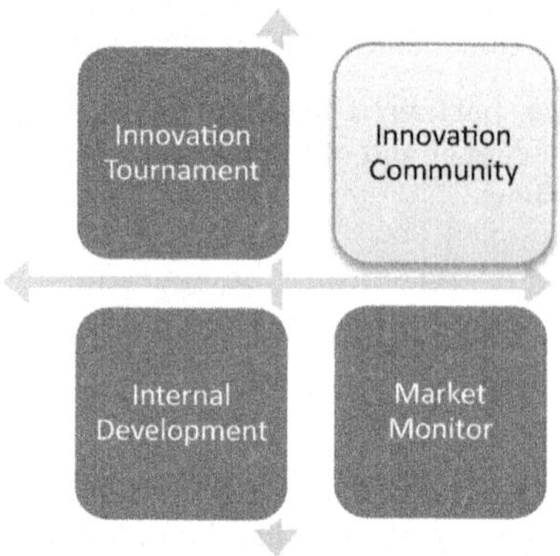

[29]M Ihlwahn (2008) Why Samsung and LG Scare Motorola. *BusinessWeek.* http://www.businessweek.com/stories/2008-05-01/why-samsung-and-lg-scare-motorolabusinessweek-business-news-stock-market-and-financial-advice.

The success of this process is perplexing- how does a company incite designers, a group that traditionally tends to sequester their intellectual property, to spontaneously divulge their secrets? In fact, the designers are honored to be associated with a company with the superior image of Alessi and the signature pieces further boost this sense of personal recognition and achievement.

Forming a consortium can lower the cost of innovation development, and also enable the firm to leverage the expertise of its partners that may be capable of developing the project with greater success or be able to devote more resources. In addition, an innovation consortium has the freedom to operate outside of corporate structure. For instance, Eli Lilly, Merck and Pfizer jointly launched the Asian Cancer Research Group, a non-profit company, to collect, analyze, and publicly share data on tissue samples from lung and gastric cancer patients for early stage research in order to stimulate cheaper and more effective innovative treatments.[30]

Organizational Zoom . . .

Google's Open Consortium

Google has engaged in a relationship of open innovation with the European Patent Office (EPO). The Internet Company has developed advanced language translation software called Patent Search to help inventors ensure their work is dissimilar to patents filed in different languages.

But what does Google stand to gain from such a relationship? Aside from being an "altruistic" supporter of invention and innovation the company has its own venture capital firm and the partnership with the EPO facilitates the streamlining and the fine-tuning of the accuracy of the software by giving Google unique access to large quantities of technical and multilingual documents. The EPO's president, Benoît Battistelli, states that "*the new translation tool is a further stepping-stone to improving innovation in Europe*".[31] Thus the "techno-altruism" of Google's support of improvements in the patenting process has

[30] http://www.ft.com/intl/cms/s/f8dc5a42-20a5-11df-9775-00144feab49a,Authorised=false.html?_i_location=http%3A%2F%2Fwww.ft.com%2Fcms%2Fs%2F0%2Ff8dc5a42-20a5-11df-9775-00144feab49a.html%3Fsiteedition%3Dintl&siteedition=intl&_i_referer=#axzz3HWSAyKV3.

[31] D Gelles (2011) Lost in translation: Tie-up with Google offers hope for cheaper patent filing. *Financial Times*. http://www.ft.com/cms/s/0/e6dcd902-8111-11e0-9360-00144feabdc0.html#axzz3IH0JEWcp.

encouraged the firm to engage in "inside-out" innovation by making key technologies available outside of the firm.

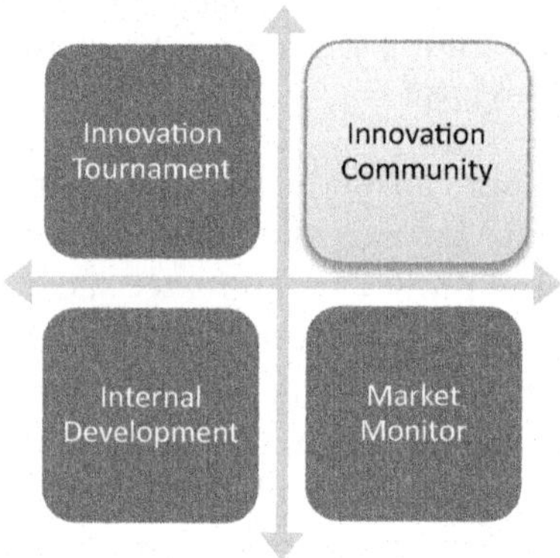

MARKET MONITOR

As Thomke and Von Hippel suggest, "the 'need' information (what the customer wants) resides with the customer, and the 'solution' information (how to satisfy those needs) lies with the manufacturer".[32] This is especially true for companies that target consumer markets, considering the complexity of modern products and services, as well as the rapidity of development and change. Hence, they need to implement a capability to sense the market preference well beyond testing and experimenting with future users or consumers.

Harnessing the Creativity and Insight of Customers

A most straightforward mindset to understand user needs and market preference is to involve customers in new product development and let them decide what they want. In this way, customers are becoming active participants, even agents of change, in the product development process. They are sources of knowledge and of information and their involvement can enhance product concept effectiveness.[33] A great example is Lego Ideas

[32]S Thomke and E von Hippel (2002) Customers as Innovators. *Harvard Business Review*, April, pp. 5–11.
[33]R Rothwell, C Freeman, A Horsley, V Jervis, A Robertson and J Crawford (1974) "SAPPHO Updated. Project SAPPHO Phase II," Research Policy, 32, pp. 58–291.

(previously known as Lego Cuusoo), a website started by Cuusoo and the Lego Group in 2008. Users can submit their product ideas to the website, and if an idea receives 10,000 supporters, it stands a chance of being produced as an official Lego product.[34] Similarly, Zara posts on the web most of the possible models for each coming season, and lets potential customers elect their favorite garments. In an effort to please the customer in this market-pull dynamic, there has been an effusive corporate interest in the notion of customer involvement. Involving the customer is considered by organizations as a way to reduce development uncertainty in the innovation process[35] and as a way to reduce the product development times or costs.[36]

Organizational Zoom ...

BMW's Customer Innovation Lab

BMW's effort began in 2003 when it posted a toolkit on its website. From the 1,000 customers who used the toolkit to develop ideas showing how the firm could take advantage of advances in telematics and in-car online services, BMW invited 15 of them to meet its engineers in Munich. Some of their ideas have reached the prototype stage. BMW is now broadening its customer-innovation efforts.

In 2004, BMW launched its Customer Innovation Lab (CIL).[37] Over 200 elaborated ideas were produced the very first year with the identification of 20 lead users. Users' ideas and concepts were evaluated by a team of BMW experts and subsequently assessed by other experts.

Organizational Zoom ...

Swarovski Mass Personalization

Swarovski launched an internet-based toolkit which enabled consumers to design their own crystal tattoos, a new generation of "body jewelry". The consumers' designs inspired Swarovski's design team and helped to analyze consumers' latent needs. Through this site more than 260 interesting motives were posted. And consumers expressed very-high buying probability of self-designed motives.

---

[34] https://ideas.lego.com/.

[35] D Leonard-Barton (1995) *Wellsprings of Knowledge.* Boston, MA: Harvard Business School Press.

[36] R Rothwell (1994) Towards the fifth-generation innovation process. *International Management Review* 11(1), pp. 7–31.

[37] https://www.bmwgroup-cocreationlab.com/cocreation/project/customer-innovation-lab.

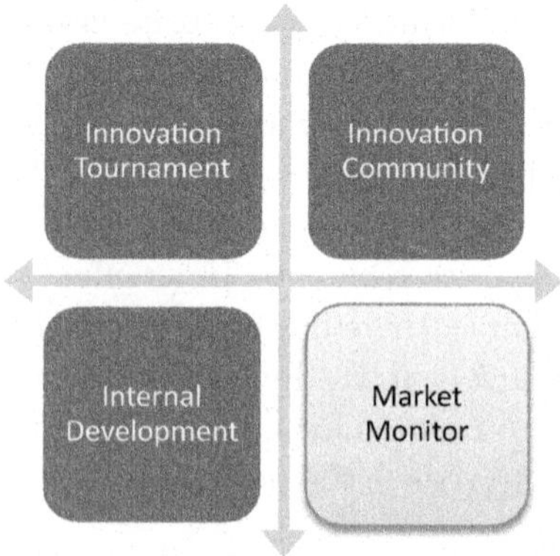

However, what accounts for the customer's motivation to contribute to the innovation process? Eric Von Hippel asked the question in 1976,[38] "*Why would anyone be so nice as to do some of your innovation work for you?*" His answer to the question of motivation is that someone will do the innovation "*if he needs the new product as much as — or more than — you do*". Customers, presumably drawn by the challenge of co-creation or motivated by the benefits of an improved product or service, are reaching out to manufacturers to tweak systems, to upgrade a program or to demonstrate to companies that very often, the customer knows best.

Indeed, end-users increasingly seek to shape a product or service to suit their needs. Companies actively facilitate such efforts of customers in participating in product and service development. Some companies have started to co-create with customers to provide customized experience. Some others are realizing that social media is an effective tool to tap the creativity of customers.

Co-Creation with Users

> "*You can't just ask customers what they want and then try to give that to them. By the time you get it built, they'll want something new*".
>
> — Steve Jobs

We are a society that has become accustomed to choice. When a straightforward black coffee in Starbucks can be recreated using a myriad

[38] EA von Hippel (1977) Has a customer already developed your next product? *Sloan Management Review* 18(2), pp. 63–67.

of alternative additions or modifications, it is no wonder that choice is more often the standard offering. Even a simplified offering such as low-cost air travel are built from decomposing the complex offering and creating a series of additional extras or "frills". David Smith, author of "*Workforce of One: Revolutionizing Talent Management through Customization*", argues that "*everything people do today in their consumer lives, from a very early age, is customized*", from ordering a Happy Meal to being able to design a pair of running shoes on Nike's Website. Once upon a time, simplicity was the norm with the Ford Model T famously available in "any color as long as it is black". Today choice and customization is seen as both a customer right, and a key mechanism for organizations to generate added value. Customers have become so comfortable with feedback that they see it as a natural inclination to propose their preferences for a product or service to the firm; the customer is unquestionably King in this 21st century sophisticated market.

Organizational Zoom . . .

Levis' Customized Jeans[39]

For more than 100 years, Levis Strauss & Co. had been the leading brand of blue jeans in the world.

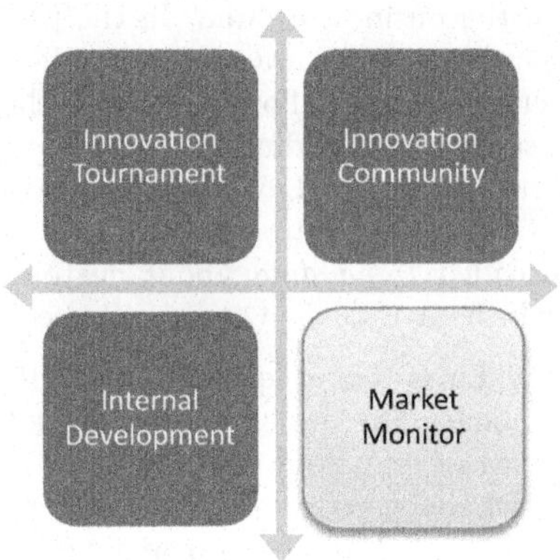

[39]http://www.cbc.ca/undertheinfluence/season-3/2014/03/01/have-it-your-way-how-mass-customization-is-changing-marketing-1/.

Then, in the 1990s, its business went sideways. It seemed that Levis had lost touch with changing tastes. While baby boomers were their loyal customers, their children were not. And with overseas factories, competitors were capable of manufacturing low-cost jeans, turning them into a commodity. Levis' 50% market share had fallen to 27% by 1997 and then 18% by 1998.

It desperately needed to innovate. So Levis decided to do something bold that no other major clothes manufacturer had ever attempted- to offer customized jeans. It was a radical strategy, as the fashion industry then was shackled to mass production by squeezing out margins through factory efficiencies and low overheads. But Levis knew that women, in particular, were frustrated with jeans shopping: 54% of them try on 10 pairs or more to find one pair they would buy; 87% wish they could find jeans that fit better; 67% felt that jeans were designed for women with "ideal" figures.

There was clearly a gap in the market. But was there a market in the gap? So Levis launched "Personal Pair" and invited women to pick the style of jean they wanted, the colour, the type of leg opening (tapered, boot cut, etc.) and the type of fly. Then three body measurements were taken. Customers could even name the jeans. The custom order was sent *via* the internet to the factory. Computers cut the pattern pieces, and the jeans were sewn manually; total time from store to customer being only two to three weeks. At that time, custom jeans cost 35% more than regular Levis. But women saw the value. And they felt the satisfaction of feeling good in jeans that were made specifically for them. As a result, over 25% of all Levis jeans sold to women became custom orders.

It helped them turn the business around. In the process, Levis learned that in the jeans category, price wars are not the only option. They learned that mass customization allowed them to charge *more* for their jeans, while at the same time building *customer loyalty* among service-hungry customers. It also allowed Levis to create a valuable database of customers when databases were rare. And best of all the custom jeans orders instantly provided *trend data* about customer preferences and in-demand styles, which gave Levis a competitive edge.

It became a strategy Levis has stuck with. In 2010, Levis introduced Curve ID: A revolutionary line of customized jeans based on the *shape* of a woman, not her waist size. And in 2012, Levis launched the "Made to Order Project", where jeans were created from scratch, by hand, for individual customers. Indeed, Levis had found a way to innovate, generate revenue, and revitalize its brand at the same time.

In many instances, the interested consumers have demonstrated an infinitely more advanced level of creativity and have improved products and services voluntarily for the company. This trend fits congruently with the greater desire and demand for customization. It is one of the chief characteristics of a more sophisticated and chameleon-like marketplace where products and services are in a continual state of development. Although many products are developed not by asking the customer, but by knowing more about the customer than he knows himself. Ignoring the feedback and input of customers can be at the peril of the market success.

Organizational Zoom . . .

LEGO: Engaging Customers in Open Innovation

"At LEGO, we stumbled across the phenomenon of customer co-creation, which is now becoming a major innovation practice".

— Jørgen Vig Knudstorp
CEO of the LEGO Group

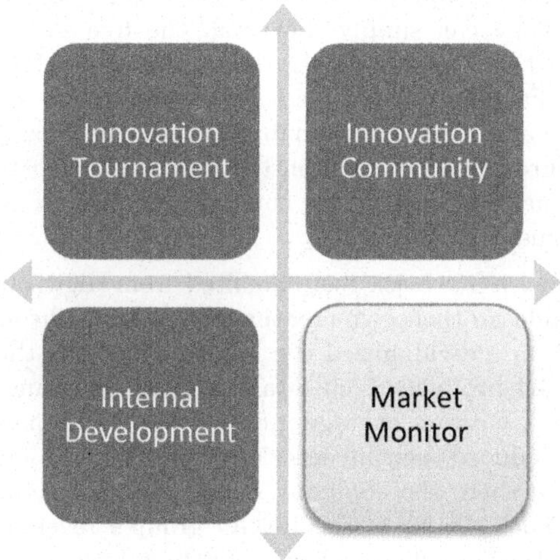

In the early 2000s, the company experienced a serious crisis with sales dropping and three years of loss. Under the leadership of the new CEO Jørgen Vig Knudstorp, thorough organizational changes and a rebuilding of the company brought the company back on track. A more loosely control system was implemented. The brand became more flexible, and the company mainly focused on attaining the customer's insights to create better experiences for all LEGO fans on current and future products. One case in point is that of Lego's development of "*Mindstorms*", the company's build-it-yourself robot development system.

At first, the company had selected four American LEGO fans to help design and innovate the new Mindstorm series. However, within three weeks of launching the system, Lego was facing around 1,000 hackers who had downloaded its operating system, vastly improved it, and posted their work freely online. It took the company a lengthy time to react to the unforeseeable event, but since then Lego appears to have accepted the merits of this community's work and swung it to the company's and consumers' advantage. Computer program written in hacker language may now be uploaded to the Mindstorms website and marketable ideas are financially rewarded.

In fact, LEGO was no longer solely engaging a community of fans in designing, developing, and marketing Mindstorms. It was now encouraging the community to evolve beyond the firm's control, doing much more than just acting as an extension of the firm's development resources. The Mindstorms community represented a whole new competence pool, in creative conjunction with employees of LEGO.[40] Using LEGO Digital Designer software, customers can now design and build any model they can imagine. They can simply download the free software, and then start designing from scratch. After that, they can upload their design to the LEGO Factory Gallery, and share with other users. Designers can choose to have their design manufactured for a fee. The customer designer's picture will be printed on the boxes of the customer-designed products. For mass production the company may also select a few of the most popular customer designs.

The traditional LEGO brick has benefited from digitization to enable customers to unleash their own creative imagination through joint value creation based on individualized experiences. In 2005, the development of sales of LEGO products from retailers to the customers turned into a positive aspect with an average increase of 9% compared with 2004. Retailers have reduced their inventories of LEGO products considerably in 2005.[41] Since then, the company experienced continuous growth in sales and revenues. In 2011, the LEGO group's revenue increased by 17%.[42] Sales of several license-based product lines were considerably above the expectations in 2011.

Now, LEGO's strategy spans product–service offerings themselves as engagement platforms, as well as engagement platforms involving traditional building kits and customized products with retailers. LEGO has

[40]http://www.venkatramaswamy.com/wiki_files/ccwiki-lego.html.
[41]Annual Report 2007, The Lego Group.
[42]Annual Report 2011, The Lego Group.

demonstrated both how to form strategic partnerships and how to open up products for user participation. As Jørgen Vig Knudstorp, CEO of the LEGO Group, notes, "We gained a better appreciation of technique and touch, and a mindset that is crucial for successful co-creation. As we have learned, co-creation is not about just 'following rules,' but personal engagement is essential".[43]

BIG Data: To Create BIG Experience

Organizations today face overwhelming amount of data, organizational complexity, rapidly changing customer behaviors, and increasingly competitive pressures. New technologies as well as rapidly growing channels and platforms have created a tremendously complex environment. The good news is that the explosion in data and digital technologies has opened up an unprecedented array of insights into customer needs and behaviors. Big Data is such game-changing opportunity for market research.

Big Data has encouraged the evolution of marketing practices. Those that use Big Data and analytics effectively show productivity rates and profitability that are 5–6% higher than those of their peers. McKinsey analysis of more than 250 engagements over five years has revealed that companies that put data at the center of the marketing and sales decisions improve their marketing returns on investments (MROI) by 15–20%.[44]

For customers, personalization is the true value of providing personal preference data. The most effective way to demonstrate personalization is to provide consumer with increasingly relevant experience, communications and offers. With more consumers shopping online and *via* mobile devices, it is easier now for retailers to personalize their interactions on the fly based on up to-the-moment data about what customers are looking at, where they are, and what devices they are using. Research shows that personalization can deliver five to eight times the ROI on marketing spend and lift sales 10% or more.[45]

[43] http://www.venkatramaswamy.com/wiki_files/ccwiki-lego.html.
[44] http://www.forbes.com/sites/mckinsey/2013/07/22/big-data-analytics-and-the-future-of-marketing-sales/.
[45] http://www.forbes.com/sites/mckinsey/2013/07/22/big-data-analytics-and-the-future-of-marketing-sales/.

Tapping the Power of Social Media

Undoubtedly, the production of knowledge is a collective phenomenon and increasingly so.[46] Especially in a world that is progressively more connected through technological innovations in social media and communications, the transfer of knowledge across boundaries is becoming a growing phenomenon. In fact, social media sites such as Facebook, Twitter, and Linkedin have facilitated the creation of a whole new communication landscape. After all, billions of Internet users worldwide are using online community sites to stay connected with their friends; make new friends and share user-created content such as blogs, photos, and videos. These rising phenomena actually present a huge opportunity for companies to interact with their customers more frequently and more easily. They also provide organizations access to a vast pool of original ideas and insights.

Indeed, social media has changed the parameters of the customer–company relationship as well as those of the traditional market push–pull diametric. In the past, firms have mostly resisted customers' innovation or insight, not known what to do with it, or remained blatantly ignorant of the potential value that it stood to generate. American farmers were lobbying manufacturers to make cars with detachable back seats as early as in 1909 and yet it took Detroit more than a decade to "invent" the pick-up truck. Nowadays, things have changed. Customers can wield the power of the social media tool in their hands, bestowing the capacity to influence the market and shape new products and services tailored to the demands of the customer. The key benefits of using social media for co-creation are that it is quick, low cost, and connects a company with a potentially enormous audience.[47] Traditional marketing channels are being surpassed in terms of influence by the channels of communication of the social media. There is a perpetual dialogue taking place on the marketplace, in which companies can choose to participate or not; either way, the dialogue will take place. Thus, many companies are taking advantage of every possible opportunity to connect with customers, for the purpose of understanding and interacting with the user, and to develop a client-centric approach to innovation.

Yet, the advantages of utilizing social media also raise another question: How to choose a social media that is best suited for your company, your

[46]RB Guimera, J Uzzi, LA Spiro and A Nunes (2005) Team assembly mechanisms determine collaboration network structure and team performance. *Science* 308, pp. 697–702.
[47]http://www.accenture-blogpodium.nl/innovation/co-creation-social-media/.

product, and your target audience? For example, if you want to build relationships with a large number of people, Facebook could be a perfect choice; if you need to tap the specialist knowledge of a group of experts, Linkedin might be more useful. In other cases, companies are attracting users to provide feedback and contribute ideas in online forums. Indeed, the rise of online communities, together with the development of powerful and easy-to-use design tools, is boosting the phenomenon of customer involvement, as well as bringing it to the attention of a wider audience.

Organizational Zoom ...

Xiaomi: Leveraging the "Fan Economy"

"Every user becomes your R&D. Every user becomes your salesman. And every user becomes your friend. That's the company we want to build".

— Lei Jun
Founder and CEO of Xiaomi

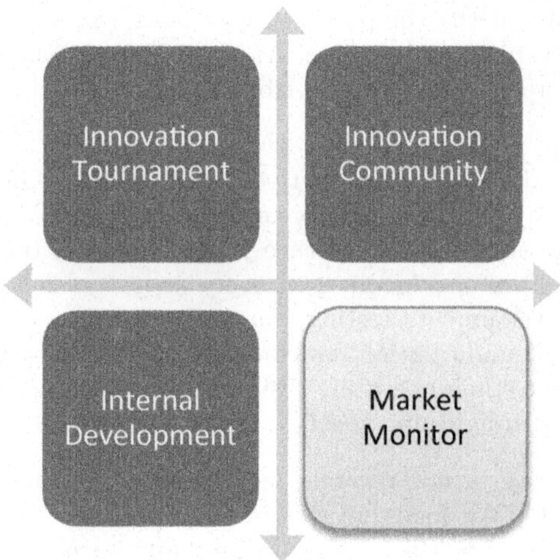

At an event touting Xiaomi's latest product release, a promotional video projected onto a big screen showed the company's young staff from the logistics, repair service and customer support departments. Different from other tech companies, which allow their engineers and designers to talk about the products they produce, Xiaomi puts its focus on its people and its users. This is the secret of Xiaomi's business success.[48]

Established four years ago, Chinese phone producer Xiaomi, being dubbed as the fastest flowing Tech Company ever, managed to catch the public's attention with its innovative and affordable cell phones. Utilizing a creative android experience and delivering high performance, the Xiaomiphone quickly gained a huge number of fans. Xiaomi generated an astounding $2.1 billion in revenue during its first full year in operation. Revenue projections for its second year are $4.5 billion.[49]

What are the secrets for this success? Fan participation and an open ecosystem for development of its phone models and operating system are the key factors. The company has taken an innovative approach to solicit and incorporate fan feedback, and this has resulted in extraordinary brand loyalty and success.

In China, as social media has developed, the term "Fan Economy", used to refer to the fact that the consumption ability of fans can be turned into cash, has become an important factor within the world of marketing. "Fan economy" is also a key term for Xiaomi's approach to marketing. For instance, the company's annual product release event, held on May 15th this year, was more than just a product presentation; it was a carnival for fans of the company. Tickets to the product release event sold out in just 10 seconds online. As Hugo Barrai, Xiaomi global vice president has previously stated, "Xiaomi carries out all its marketing *via* social media".

Xiaomi has cultivated a "hyperactive" forum, where fans can interact with the company and each other. Now the forum has some eight million users, who give the company feature suggestions, file bug reports, and user feedback. Fans are involved in every aspect of the product, from research to development, testing, marketing, and even PR; and are considered the foundation of the company. Also, the company release weekly updates to their operating systems, incorporating fan suggestions and feedback with lightning speed.

"Xiaomi's priority is not revenue, not profit, nor market share," Lei Jun has stated, "We focus on making the product that makes users scream". In achieving this goal, Xiaomi's approach is innovative and effective: One with a bottom-up dynamics whereby a demanding, rabid fan base co-creates the product in lockstep with the company. Xiaomi has disrupted the mobile phone market in China, putting pressure on Apple, Samsung, and other manufacturers. The company has also begun

[48]http://english.cri.cn/12394/2014/06/30/2701s833895.htm.
[49]http://www.innovativechina.com/2013/06/xiaomi-phones-a-co-created-phenomenon/.

selling phones in Singapore and plans to enter the Southeast Asian market.

Open innovation is a business strategy to advance technology by fusing external ideas, products, and services with internal R&D. It is a way for businesses to minimize costs and create products more efficiently. However, choosing the right path can be difficult, as organizations must have a solid understanding of their company's capabilities, and at the same time carefully evaluate the advantages and disadvantages of using outside help against developing technology internally. Indeed, many companies have embraced open innovation today, but a lot of them are still experimenting with the right approach and the mix of internal and external resources. To decide whether to engage with external partners and what kind of collaboration to be employed, companies should begin by assessing their existing knowledge and capabilities. Companies must determine if they are in the best position to anticipate the consumption of future product or service, as well as if they have the required solution knowledge and development capability to develop and implement ideas. Depending on a company's needs and resources, suitable open innovation approach would be employed and integrated into a business strategy.

Chapter 6

ORGANIZATIONAL STRUCTURES THAT STIMULATE INNOVATION

Today, the resources and capabilities available to companies for product and service development are more widely shared, but they are accordingly complex to navigate and leverage successfully. Today, the possibilities that exist for leveraging innovation are immense — entire networks of users, suppliers, and technology lie at the doorstep of the enterprise waiting to be utilized. Not only do companies have to look within the walls of the organization to see where creativity and innovation can be leveraged, they also have to look at the entire ecosystem and value chain to transform the resources of the ecosystem into valuable innovations.

Yet, the fast pace of the innovation-driven economy has left many companies disorientated within the maze of the diverse and rapidly shifting competitive environment. Companies are stalled by the challenge of transforming the structure of the organization into a culture that is service-driven and favors idea-generation. Many companies are left at a standstill, unable to react to the changing needs of users and new competition and are left to dwindle with a product or service offering that may once have seemed like an infallible source of profit. The only road to sustainability is finding the right formula to leverage the innovation challenges successfully. While the possibilities for innovation remain immense, effectively managing the tools and networks to leverage these possibilities remain largely un-mastered by companies. Companies such as Apple, Samsung, and Microsoft have succeeded in building the right capabilities and culture for innovating and utilizing the resources that lie within and outside the company;

however, maintaining this advantage is another wall to climb, which requires continued revitalization, and reinvigoration; companies simply cannot afford to sit on their laurels basking in their present success. Tomorrow may pose an entirely new set of challenges.

A 2012 McKinsey survey on innovation unveils that nearly two-thirds of companies report broad innovation portfolios that include more than one type of organizational model and nearly half use separate innovation functions that focus on developing new business opportunities.[1] 86% of the executives report that the structure of their separate functions positively influences outcomes. However, the most important factors for success seem to be the extent to which innovation is integrated in corporate strategy. While innovation department now usually take responsibility for new ideas to prototype and report in-market introduction, there doesn't appear to be a consensus on the ideal organizational model. For instance, only 4% of the companies surveyed by McKinsey depend on innovation incubators, which commonly emerged ten years earlier.

INNOVATIVE ORGANIZATIONAL STRUCTURES

Dominated by the "Taylorist" logic of rationalization and efficiency, the transition from the industrial era[2] to the current day model which is characterized by the increasing value of knowledge and services, has changed the rules of business dynamics. The creativity required for today's organizations marks a departure from the conformist, traditions of conventional management processes where authority was revered and respected, and acceptance to the organization was predominantly on the basis of fitting in and conforming to the typeset character of the organization. Clearly, traditional corporate structures are being challenged by the growth of new companies that uniquely structure their operations to meet rapidly changing customer demands.

Nowadays when it comes to product and service innovation, the most successful companies are those whose organizational structures foster the development of knowledge through formal research and development

[1] http://www.mckinsey.com/insights/innovation/making_innovation_structures_work_mckinsey_global_survey_results.

[2] R Florida (2002) *The Rise of the Creative Class: And how it's Transforming Work, Leisure, Community and Everyday Life.* New York: Perseus Book Group.

processes and the development of knowledge based on experience, practice, and interaction between employees, clients, and suppliers.[3] Coordination, collaboration, outsourcing and open-source software have been keys that have helped shape new business structures. While centralized authority was the driving force of business in the 20th Century, brainstorming and independent business divisions characterize emerging companies of the new century. Innovative organizational structure is less formal and more likely to allow creative inputs from teams and individuals.

Today, we observe a greater degree of creativity in the structure and the management approaches of organizations in responding to the needs and challenges of today's business environment. For instance, practices such as "free time" which are practiced in Google and 3M, allow employees to pursue creative ideas for innovation in their own time. Diversity and divergence of opinion, and not conformism are the principle sources of value which enable a broader perspective on things. The "scientific management" approach of Frederick Taylor, which was outstanding in its contribution to modern management practices and organizational efficiency, is inadequate for the modern working environment in which a greater degree of creative freedom for all employees is growing in importance. This quality is not restricted to purely "creative" industries and knowledge-based activities. A case in point is the success of the Toyota Total Quality Management (TQM) model, which is based on the premise that individual contributions and propositions are of utmost importance and contribute collectively to the greater good of the organization. In this "empowered" system, individual roles are infused with meaning and employees are respected for the creative solutions they bring to the organization, as opposed to being treated as an insignificant "cog in the machine".

Organizations can be structured in various ways, and the structure of an organization determines how it operates and performs. To constantly stimulate innovations in the organization, it is important to give opportunity to every employee, and to customers, to express themselves in the best possible way. Specifically, companies currently adopt four common types of organizational structures: Multidivisional structure, program structure, project structure, and matrix structure.

[3]MB Jensen, B Johnson, E Lorenz and BA Lundval (2007) Forms of knowledge and modes of innovation. *Research Policy* 36(5), pp. 680–693.

Multidivisional Structure

A multidivisional structure groups each organizational function into a division, based on products, services, or geographic locations. Each division contains all the necessary resources and functions such as its own finance, IT and marketing departments within it to support that product line, service, or geography (see Figure 6.1). For instance, a manufacturing company might implement a product-based divisional structure that includes the cable assembly and product engineering divisions, while a professional services firm might organize around service lines, such as the personal and business services divisions. As an alternative, fast food chains that have similar branched facilities located in widely dispersed areas might create a geographic-divisional structure, to cater to the particular needs and preferences of customers in different geographic areas.

Multidivisional structure, also known as M-Form, is useful to companies who compete on the global market. One parent company owns subsidiary companies, each of which uses the parent company's brand and name. For example, Virgin Group is the parent company of Virgin Mobile and Virgin Records. In some of these organizations, the parent company simply provides funding to its divisions or independent profit centers, while letting those entities have autonomy for developing their own business strategies. Many big companies like Johnson & Johnson, GE and AT&T are broken down into separate divisions headed by presidents, and allow their divisional executives freedom to manage their operations independently.

A major benefit to this structure is that each division is sensitive to the local changing customer preferences and is thus able to develop and produce

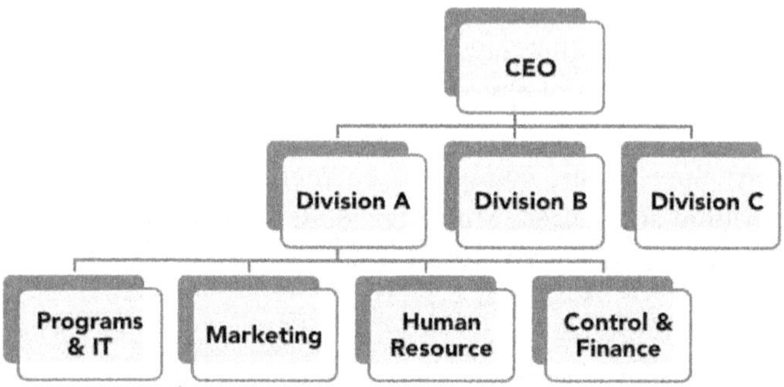

Figure 6.1: Multidivisional Structure.

goods and services that appeal to its market. Besides, each division produces and transports using local resources. In addition, senior managers assume more responsibilities for the results of local operations that are within their control. This structure can be useful because it affords the company greater operational flexibility. Disadvantages of this structure include duplication of functions at each division, and lack of standardization and operational inefficiencies in separating specialized functions.

Organizational Zoom ...

Procter & Gamble (P&G)'s Multidivisional Structure: Thinking Globally, Acting Locally

Consumer giant P&G has made its organizational structure an important part of its capability to grow. The company's organizational structure is made up of Global Business Unit (GBUs), Global Operations, Global Business Services (GBS), and Corporate Functions (see Figure 6.2).

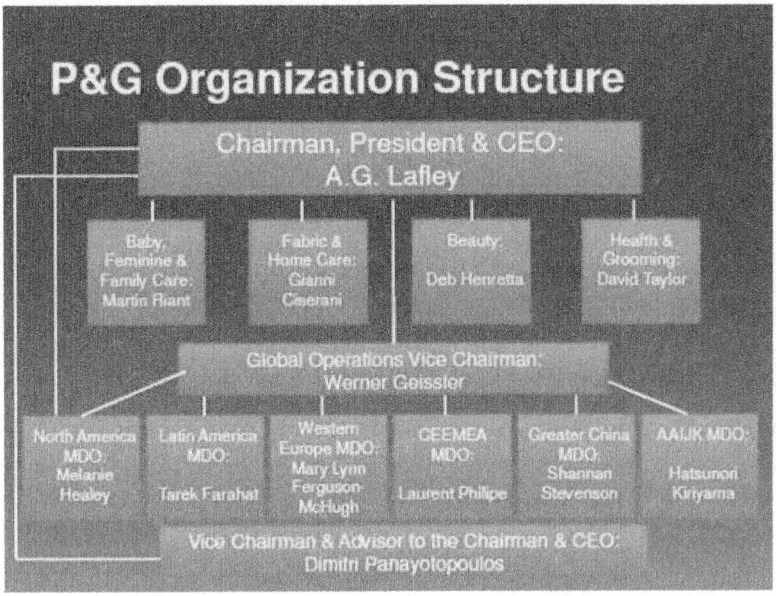

Figure 6.2: P&G's Multidivisional Structure.

Source: Deutsche Bank Global Consumer Conference[4] (Fiscal year 2014).

[4]http://www.pginvestor.com/calendar.aspx?iid=4004124.

P&G's most recent organizational chart shows that the GBUs are organized into four industry-based sectors: Global baby, feminine and family care, global beauty, global health and grooming, and global fabric and home care (see Figure 6.2). Each GBU is responsible for a broad category of products, the overall strategy, R&D, industrial production and the upstream brand marketing. They ensure that P&G put on the market the best innovations at the earliest ("speed-to-market").[5]

Global operations run the Selling and Market Operations (SMOs) and are organized geographically [includes Asia, Europe, India, the Middle East and Africa (IMEA), Latin America, and North America].[6] They are responsible for the local business operations and implementation of marketing plans, including selling, distribution, shelving, pricing execution and merchandising. They have direct contact with consumers, distributors and other stakeholders, and ensure P&C wins the "first moment of truth". The GBS utilizes P&G talent and expert partners to provide business logistics and administrative support to the GBUs and SMOs, taking advantage of P&G's scale to save on costs. Lean Corporate Functions ensure ongoing functional innovation and capability improvement.[7]

Indeed, such a cooperative multidivisional structure allows P&G to "think globally and act locally". The company thinks globally by leveraging the four GBUs to create strong brand equities through ongoing innovation, while it acts locally by using SMOs to interface with customers to ensure that a division's marketing plans fully capitalize on local opportunities. In addition, the company uses GBS group to support operations through infrastructure services such as HR and accounting, and Corporate Functions to assure that the latest and most effective methodologies are being used to conduct the firms product- and marketing-oriented operations.[8] Such structure is believed to be a key source of competitive advantage for the company.

Program Structure

A program usually groups a series of projects on sub-projects. A set of associates with complementary skills are assembled and share a common objective while completing individual tasks. A variety of expertise are put together and teams address specific operational tasks. Cooperation is then critical in program–based organizations, and this type of organizations

[5] http://www.pg.com/fr_FR/company/global-organization.shtml.
[6] http://www.pg.com/en_US/company/global_structure_operations/corporate_structure.shtml.
[7] *ibid.*
[8] http://www.swlearning.com/management/hitt/sm6e/hitt_sm6e_338_371_ch11.pdf.

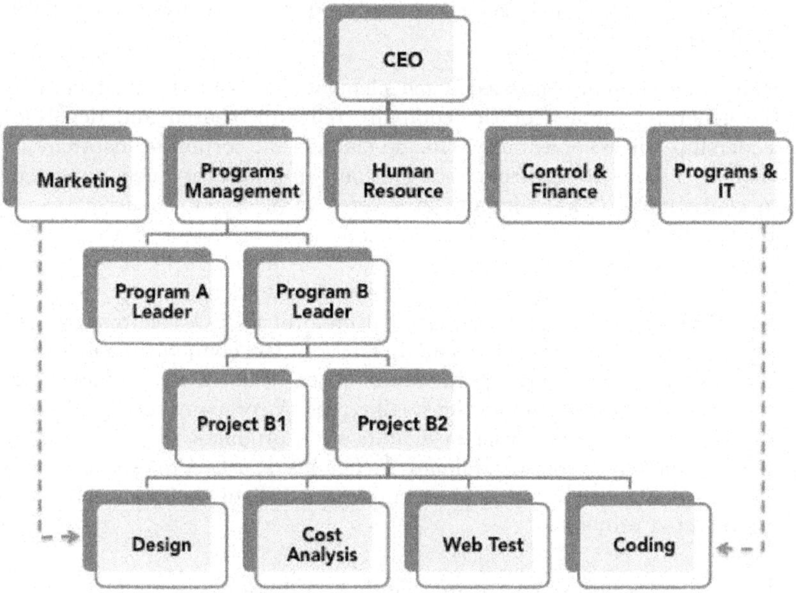

Figure 6.3: Program Structure.

should be structured to foster communication. Teams may change and adapt to fulfill group and organizational goals. Some teams endure over time, while others such as project teams are disbanded when the project ends. See Figure 6.3 for details.

Organizational Zoom . . .

Google's Innovative Organizational Structure

It is hard to think of a company that has experienced more success in the past decade than Google. In 1998, Google was founded and since then it has become the most utilized search engine in sixteen countries, based on the number of unique visitors to the site. What started out as a student project by two Stanford graduates turned into one of the most recognizable companies in the World!

The two masterminds behind the relatively simple, yet very profitable idea of the Google search engine were the founders of the company, Larry Page and Sergey Brin. At the beginning, they employed an organizational structure based on a traditional hierarchy; yet, they found the structure to be too top heavy and clumsy. In 2001, Larry Page and Sergey Brin decided to reorganize the company and implement a decentralized, team-based structure. The company now has a very flat level of hierarchy

> in which top-level management is only one step up from lower level employees.
>
> While the company utilized some elements of hierarchical structure, the general approach of the company relies on shared and devolved leadership and responsibility. Google claims that corporate hierarchy is virtually non-existent throughout the company and everyone is expected to contribute to the company in several ways. Small, creative teams are set up for individual projects, and project leaders of the teams change with nearly every new project. The members of these teams are also quite different because of the diverse backgrounds of Google's employees. Google's employees range from former CEOs, neurosurgeons, wrestlers, and even marines. The diversity of the company's employees is what helps the company in building new innovative products and interacting so successfully in the small teams. A company such as Google strives on coming out with new innovative products and the owners of the company were so adamant in helping spark employee product development that they gave their employees 20% of their work time on self-directed projects.

The team-based structure in large organizations is a relatively newer type of organizational structure. Unlike traditionally structured organizations that rely on top management to control all work activities, team-based organizations filter decision making down to all levels of management. Team-based organizations are less hierarchical, less structured, and more fluid than traditional structures (such as functional or divisional). In such organizations, employees participate in problem-solving, decision-making, and teamwork to achieve organizational goal. To a certain extent, when the pharmaceutical group Pfizer Inc. founded its Cambridge, Massachusetts-based Research & Technology Center in 1999, the company was looking for a novel way to address its development effort. It was the smallest, but also the most expensive R&D facility on a per capita basis. With the firm eight units R&D global deployment by the mid-2000s, RTC was established as an experimental site.[9] The site, originally termed The Discovery Technology Center, provided and fostered a biotech-like atmosphere, consistent with its role as an agent of change.

A team-based structure enables a company to capitalize on employees' strengths and stimulate creativity and innovation. Software developers, for example, may divide a project into teams of individuals who each contribute

[9] S Thomke and A Nimgade (2008) Pfizer Inc.: Building an innovation center. Harvard Business School case series No. 9-609-037.

their specialty without necessarily reporting to a manager, although major corporations typically do have team supervisors. Team diversity which pools hundreds of innovative minds together can lead to "out of the box" ideas that may have otherwise been overlooked by executives who cling to unchallenged company traditions. In addition, teams can provide a unified spirit to a company such as Starbucks scheduling cross-country team field trips to its restaurants, inspiring staff to learn from the customer experience.

Organizational Zoom . . .

BMW's "Chaotic" Collaboration

Innovative "lateral" management practices lie behind the success of the buzzing collaborative innovation culture that permeates the BMW offices. As BMW chief designer Adrian van Hooydonk notes, "*At lunch and breaks everyone is discussing ideas and projects all the time. It's somewhat manic. But it makes things move faster*". The informal networks that flourish at BMW create a seemingly chaotic environment, but it is precisely this "chaotic" collaboration through informal lines of communication that contribute to the highly innovative environment. But how has BMW managed to foster such unbridled collaboration?

BMW has taken an unconventional approach to management hierarchies-cross-functional networks, as opposed to hierarchies. This innovative approach of organizing people dominates daily operations and enriches working methods. From their first day at BMW, employees are encouraged to forge a network of personal ties to speedup problem-solving and innovation, be it in R&D, design, production, or marketing. The networks run laterally across divisions as well as vertically up and down the chain of command.

According to CFO Stefan Krause, the human network approach is of key importance: "*Good companies have this lateral ability to communicate across divisions and silos, not just up and down the hierarchy. That's what makes BMW tick*". Workers are free to build teams across departments or divisions, to seek out the internal talents that tend to be disconnected in most companies. This optimization of talent and collaborative developments helps employees meet their targets in a more effective way.

The boundary-less culture that favors fresh thinking and unconventional ideas goes against the grain of traditional German corporate culture and the according psyche. Yet, people soon learn that seditious attitudes are preferred, and as one worker remarked, "*It's easier to ask forgiveness for breaking the rules than to seek permission*". Workers don't hoard information from one another, they take the time to work together; instead of an individualist corporate environment, a collective culture reigns which are more open to change and where risk is accepted are key components of the innovative culture. The unique innovation culture has led to a number of very successful innovations across all departments that come down to the free reign of ideas in the company.

Project-Based Structure

A project is an organizational effort with a specific objective that needs resources, and must meet time, cost, and quality requirements. A variety of businesses are naturally project-based, such as construction, entertainment, aerospace, and so on. In a project-based organization, most of the business functions are organized in projects unlike a functional structure, whereby work is organized in departments based on specializations such as marketing and finance. Usually, there is a general manager or CEO; also there may be functional departments such as operations and finance, but projects are the main units of conducting activities. Project teams conduct almost all operations of the company. Projects are focused on solutions and implementations, such as whether the project achieves its target, whether it is within budget, and whether the team will complete the task on schedule. Functional managers make sure that projects have access to employees with the right functional expertise. Project managers have a high level of authority to manage and control the project resources, and are responsible for delivering results. They select team members, allocate tasks, and oversee performances (see Figure 6.4).

Companies often set up projects to explore new business opportunities. A project involves the development of a specific product targeted at select customers. A big company engaged in continuous projects can enhance its culture by bringing different department leaders together in one physical environment. BMW coordinates its innovative car design projects by relocating members to its research center, FIZ, for three-year periods. Engineers and marketers work together in person, which accelerates communication and reduces late-stage conflicts. Known as "cross functional collaboration", this brainstorming technique can speed up project timelines from years to months. Some companies work on multiple projects at one time so that leaders can select the most promising opportunities for product releases.

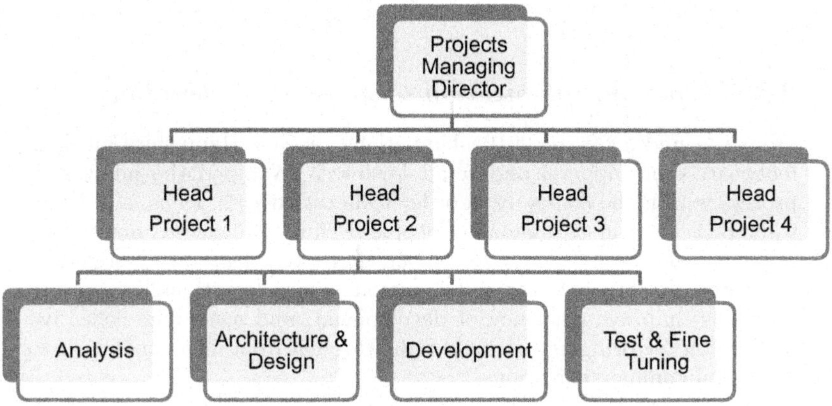

Figure 6.4: Project-Based Structure.

In the project-based organizational structure, employees are specifically assigned to projects and report directly to the project manager. The project manager conducts performance appraisals and prepares career progression of all project team members. This leads to increased employee loyalty to project goals. In addition, the project manager has direct control and authority over project operations, which result in improved reaction time to changing circumstances and flexibility in assigning work to the employees best suited for the tasks. For example, sales people are normally part of the sales department in a functionally organized business. In a project-based business, a project manager can put them in a project to work with engineers and operating personnel to find out how the products they sell are designed and produced. They could be then assigned to a sales-oriented project to apply their new knowledge to effective use.

Pure project-based organizations are more common among large and complicated projects. These large projects can absorb the cost of maintaining an organization whose structure has some duplication of effort and the less than cost-efficient use of resources. In fact, one major disadvantage of the project-based organization is the costly and inefficient use of personnel. Project team members are generally dedicated to one project at a time, even though they may rarely be needed on a full-time basis over the life cycle of the project. Project managers may tend to retain their key personnel long after the work is completed, preventing their contribution to other projects and their professional development.

Organizational Zoom . . .

Timberland: Reorganization to Foster Collaboration

From the mid-2000s to 2010, Timberland, a New Hampshire outdoor footwear, sport apparel and retail business, revamped the innovation process within the company in order to boost success, focus, and time-to-market. Instead of organizing people by function such as marketing, R&D or sales, Timberland tested an alternative method of organization that sought to foster collaboration across traditionally separate functions, improve efficiency of development, and encourage collective motivation for bringing products right through from idea generation to successful commercialization.

Cross-functional groups and the *In-line* teams were organized by customer segments such as women's casuals, kids wear, etc. This reorganization not only enabled collaboration between diverse profiles and collective expertise, but it also generated an acute and intimate knowledge of the particular customer segment by enabling the teams to really get to know their market. The result was an improvement in gross margins and an optimization of efficiencies. However, multi-functional product teams focusing on market segments experienced a tension between function and fashion.

To address these conflicts while fostering innovation, the company divided marketing and design groups into three teams: In-line, as described above; speed, in charge of creating rapid design adjustments to existing products; and advanced concept, known as Invention factory (iF).[10] First iF initiative called Travel Gear, encountered difficulties getting wholesaler acceptance, despite a successful pilot with retailers. As a consequence, iF's missions were redefined in order to focus on the relationships with the in-line teams since the difficulty in handing off new products to the operational teams was perceived as the main issue.

Matrix Structure

A matrix organization structure is defined as one where there is a dual structure combining both functional specialization and business product or project specialization.[11] Usually there are multiple reporting lines, that is, people have more than one formal boss. This may include multiple direct

[10]RM Kanter and RL Raffaelli (2008) Innovation at Timberland: Thinking outside the shoe box. Harvard Business School case series No. 9-306-064.
[11]JG March (1994) *A Primer on Decision Making: How Decisions Happen*. New York: The Free Press.

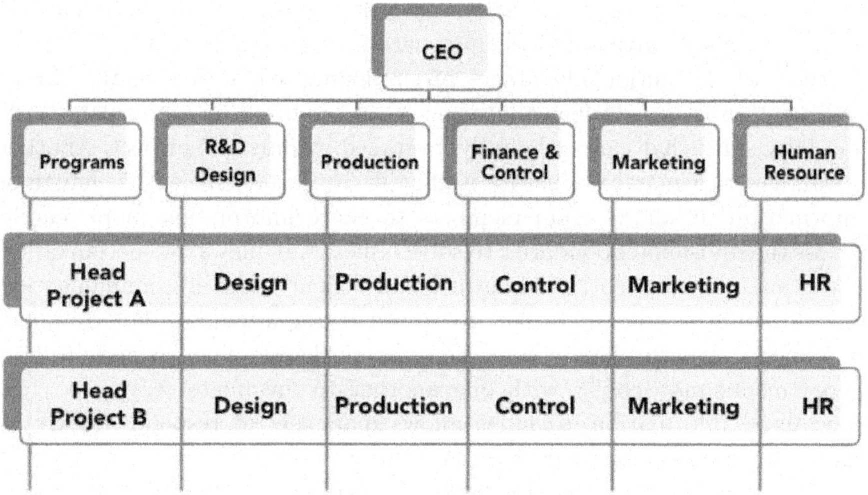

Figure 6.5: Matrix Structure.

strong reporting relationships to more than one boss, or a strong reporting line as well as a weaker reporting relationship that still indicates some formal level of "right" to the individual's time (see Figure 6.5). A matrix organization allows its program units to focus on their specific technical competencies and allows projects to be staffed with experts across the organization. For example, a health specialist might report to the health division, but be temporarily allocated to work on various projects. Although complicated, an effective matrix structure can lead to better coordination among a company's divisions.[12]

The matrix organizational structure brings together employees and managers from different departments to work toward accomplishing a common goal. Project managers who bring together specialists from within the company and beyond to work on multiple projects operate under a Matrix Structure. The concept can be extended to include outsourcing specialists, similar to what P&G did when they reorganized their research department, embracing collective ideas from around the world to generate new products. By interfacing with networks from the scientific community, the company has created new job positions such as "technology entrepreneurs", who scout universities and other labs for new scientific discoveries. Magnetic

[12]P Walter (2003) Executive agenda column. *Bangkok Post*, May 1st. http://www.proquest.umi.com.

tape manufacturer 3M along with many other technology companies, award grants to outside inventors to help generate innovative products.

One of the major advantages of employing a matrix structure in an organization is the efficient allocation of all resources, in particular scarce specialty skills that cannot be fully leveraged by only one project. Another advantage of a matrix structure is that it can foster efficient exchange of information. It allows team members to share information more readily across the divisional boundaries to solve issues, and allows for specialization that can increase depth of knowledge. Efficient lines of communication enhance productivity and allow for quick decision-making. For example, in a matrix structure, individuals from the marketing, finance, and product departments may confer with one another to formulate strategies. The specialized information exchange allows managers to respond quickly to the needs of customers and the organization.

The main disadvantage of the matrix structure is that it could result in internal complexity. Some people might report to division managers for whom little work is done, while working for one or more project managers at the same time. It becomes more important for employees to develop strong time management skills to ensure that they fulfill the work expectations of multiple managers. In addition, such organizational structure requires effective communication and cooperation between multiple divisions and project managers, as they are all competing for scarce resources such as time and talent. Otherwise, miscommunication and ineffective managing can result in employee dissatisfaction and low morale.

Organizational Zoom . . .

Reshaping Cisco: A Matrix Approach

An Internet technology company Cisco Systems, Inc. (Cisco) which has reinvented itself time and again has proved that the key to corporate success lies in an organizational structure that is both responsive and in tune with the changing industry and market requirements.

Earlier in 2001, the company reorganized itself, forming cross-functional teams to break free of the "silo culture" earlier prevalent in the company. Over the years, it refined the model and came up with an organizational structure combining a functional structure with cross-functional groups, called a "matrix". This structure was put in place to support Cisco's transition from being just a seller of routers, switches, and other technology to being a company that is the most trusted business and technology adviser to its clients.

It has developed an elaborate system of committees made up of managers from different functions which is different from those companies that are organized based on lines of business. Essentially, the key decisions in the company are taken by councils, boards, and working groups. The job of most of these groups is to tackle new markets. "Councils" are in charge of markets that could reach $10 billion. For "boards" this number is $1 billion. Both are supported by "working groups", which are created as needed. There are about 50 boards and councils, with some 750 members. Cisco has given up counting the working groups, because very quickly they come and go.[13]

To make this seemingly complex structure work, efficient communication and coordination across groups and functions are of great importance. The firm developed a culture of collaboration from the top down. John Chambers, the chairman and CEO of Cisco, cultivates a co-operative management style. Some council does without a formal leader and function more like a sports team. Many managers have leading roles both in a function and on a council/board, which fosters co-operation. How well managers do in teams determines 30% of their bonuses. There have been casualties: Whereas those who work well with others have been promoted, lone fighters have been pushed out. As a result, a fifth of Cisco's leadership has left the company.

In addition, the company developed "replicable processes": Jargon for rules covering how the groups are set up, how their work is evaluated and how decisions are taken. There is also a "common language" in which groups must describe their work. Each has to come up with a statement that includes a five-year vision, a two-year strategy and a ten-point execution plan. This not only imposes discipline but also makes decisions transparent.

According to Mr. Chambers, a matrix structure can help the firm to react quickly to new opportunities, and also makes it much easier to come up with entire solutions rather than stand-alone products.[14] This is what many customers, particularly governments, now demand. And the structure helps Cisco to become a globally integrated company by making it easier for executives from all around the world to weigh in.

Some think it could even become a model for the corporation of the future. Rosabeth Moss Kanter, a professor at Harvard Business School, regards Cisco as an example of a "supercorp", a coinage that is also the title of one of her books. It has avoided becoming a lumbering and bloated giant which is the fate of many other companies as they grow. Tom Malone, a professor at the MIT Sloan School of Management, sees Cisco as a pioneer for a larger trend. Traditionally, he says,

---

[13]The Economist (2009) Reshaping Cisco: The World according to Chambers. Aug 27th.
[14]*ibid.*

> management was about "command and control". Now, as technology makes communication much cheaper, bosses should move to a more flexible view, best described as "co-ordinate and cultivate".

Ambidextrous Organization

Organizations designed for large-scale production and delivery are typically poor parents for internal new ventures. At many large organizations, when faced with the resource constraints, functional managers are often reluctant to allocate funding and staff to projects they perceive as risky, or those that don't help them achieve their short-term performance metrics. In addition, inflexible infrastructure and support structures, and inconsistent sponsorship are also symptoms of many large organizations. Established enterprises are inherently designed using metrics, processes, and reward systems suitable for mature businesses, not for fast-moving start-ups.[15]

Ambidextrous organization designs create distinct units that have their own unique processes, structures, and cultures that are specifically intended to support early-stage innovation. These units which are often comprised of one or more innovation teams reside within the larger parent organization, either spun out as new business unit or folded back into existing units. New opportunities are identified, developed, and brought to market by these units. They are set up to support the unique approaches, activities, and behaviors required for launching a new business.

Companies can use the ambidextrous organizational structure to create segregated business units for exploring and developing breakthrough innovations (products, services, and processes) while at the same time keeping existing business units intact. Project teams within the new venture are encouraged to form their own processes, structures, and cultures but they are still connected to the rest of the organization through executive sponsors who ensure that no organizational conflicts or competition for resources threaten the viability of the venture.

To develop an ambidextrous organization, leadership must possess the ability to attend to already existing products, services, business models, and processes while simultaneously supporting the innovations that will drive the organization's future. Ambidextrous design and management ultimately means continuing to oversee the core business while concurrently protecting the emerging venture as it evolves and grows.

[15]http://www.innovation-point.com/ambidextrous.htm.

Organizational Zoom . . .

Siemens R&D Organizational Structure

Siemens is one of the world's largest firms, operating in the industry in energy, healthcare, and infrastructure & cities sectors. Its 370,000 employees generated annual revenues of USD 102 Billion in 2012.

At Siemens, R&D activities were conducted by Divisions, as well as by Corporate Technology (CT). The businesses focus their R&D efforts on the next generations of their products and solutions. In contrast, the aim of CT is to be a strong innovation partner for operational units and to help secure their technology and innovation future while strengthening Siemens as a technology company.[16]

Siemens' exploratory unit, CT is the competence center for technology and innovation of the firm and an integral part of its global knowledge network. CT can support development work from initial idea to pilot product or technology. It works to safeguard the company's technological lead and boost its competitiveness.

One particularly interesting aspect of CTs work is the approach and the energy invested in communicating breakthrough innovations and innovation insights. CT has even installed an organizational unit responsible for communicating the exploratory work.

One focal point of communication is CTs "lighthouse projects" which are publicly communicated. Lighthouse projects are strategically important projects that have the potential to open up new fields of business *via* group-breaking technologies. They combine researchers from various clusters and sectors in order to develop solutions that would radically change the business for their respective operative units. The lighthouse projects send a clear signal about the long-innovation focus areas into the organization and to the outside world.

A second focal point of communication is the "*Picture of the Future*" magazine. This describes in-depth sophisticated future studies, major technology trends and radical innovations developed by Siemens which are well understandable even for non-subject experts. The "*Picture of the Future*" is also distributed outside of Siemens to support the innovation claim of the corporate brand. Since 2001, the magazine is distributed twice a year.

We have reviewed several organizational structures in this section. Without a specific context, no organizational structure is inherently

[16]Siemens company website. http://www.siemens.com/innovation/en/about-research-and-development.htm.

superior to the other structure.[17] In Peter Drucker's words, "There is no one right organization. Rather, the task is to select the organization for the particular task and mission at hand".[18] When exploring potential organizational structures, the company must understand what models currently exist within the organization, and then evaluate the extent to which these various structures are aligned with and support the company's strategic intent and strategic mission. Since no single structure is optimal in all instances, managers should concentrate on developing proper matches between strategies and organizational structures rather than searching for an "optimal" structure.

ORGANIZING FOR OPEN INNOVATION

Innovation has traditionally been solely located in the realm of companies' internal activities, such as in-house R&D. With the movement toward "open innovation", firms increasingly open up their boundaries in order to access external sources of knowledge and technology or bring in-house inventions to markets *via* external paths.[19] Yet, open innovation cannot be implemented in organizations without the right organizational structure supporting it. Business leaders now recognize that entirely new strategies and approaches are required; not just market-driven strategies but organizational strategies, in order to drive open innovation and sustainable growth. Even with such an appreciation for innovation, a big question often remains: How do you formalize an organizational capability for identifying, evaluating, selecting, and commercializing the most promising opportunities for the company?

Internal vs. External R&D

R&D activities take place in various organizational forms depending on who finances, creates, develops, produces, and sells the innovation. A widely observed organizational form is in-house R&D. The innovation is financed

[17]B Keats and HM O'Neil (2005) Organizational structure: Looking through a strategy lens. In MA Hitt, RE Freeman and JS Harrison (eds.). *The Blackwell Hand-Book of Strategic Management.* Malden, MA: Blackwell Publishing Inc.
[18]MF Wolff (1999) In the organization of the future, competitive advantage will be inspired. *Research Technology Management* 42(4), pp. 2–4.
[19]C Ihl, F Piller and P Wagner P (2012) Organizing for open innovation: Aligning internal structure and external knowledge sourcing. *Paper presented at the DRUID 2012.* CBS, Copenhagen, June 19th.

and produced within the same firm who uses the new product or the new technology. Many large firms have restructured their R&D processes within firms to improve the linkages between their R&D programs and their overall strategic objectives. The implications of these changes are perhaps most pronounced in centralized corporate research labs, which perform the most fundamental research in the business sector.

Firms have experimented with a range of reforms that strive to provide researchers in such labs with sufficient freedom to explore new scientific and technological opportunities with uncertain outcomes while ensuring that the labs contribute to the profitability. Elements of the restructuring include[20]:

- New funding models. Funding of internal research laboratories relies less on central funding and more on mixed models in which researchers receive some financial support from product divisions.
- Links to the market. More explicitly links are established between research programs and market needs, whether by researchers working more closely with customers or through more elaborate research planning processes.
- Re-organizing staff. Organizational structures based on traditional academic disciplines are being replaced by problem- or product-oriented structures. Incentive plans are rewarding researchers and research managers for both quality research and contributions to the business performance.
- Leveraging intellectual property. Firms are increasingly patenting inventions and licensing that do not match their corporate objectives to other firms, in some cases, even spinning off separate companies to commercialize the technology. The revenues from licensing streams (or from the appreciation of stocks in spin-offs) can provide a financial return to technology that might otherwise go unused within the firm.

Besides in-house R&D, innovations can also be produced externally. In fact, for many firms, a significant aspect of the restructuring of R&D operations has been a conscious attempt to open up their R&D systems to better integrate with external sources of technology. The objective of this approach is to increase the flow of ideas through the company, both to make researchers aware of external developments of interest to the firm

[20]HW Chesbrough and SJ Socolof (2000) Creating new ventures from Bell Labs technologies. *Research Technology Management* 43(2), pp. 13–17.

and to speed up subsequent innovation processes. This approach may allow firms to outsource some of their R&D activities, but more importantly, it can enable them to take advantage of a wider range of technological inputs.

Firms report that they increasingly outsource basic research to R&D service organizations and more frequently partner with universities and national laboratories to develop technology.[21] In addition, small technology-based firms (e.g., high-tech start-ups) have also become a growing source of new technology. Such firms concentrate more on markets that are initially small, and are thus often more effective than large companies at commercializing radical innovations that open new product markets.

While large firms finance some R&D in small firms and license the results of such work, other mechanisms, such as corporate venture capital (CVC) funds and mergers and acquisitions (M&As) have become increasingly popular for financing and appropriating the results of start-up R&D:

- Development activities are conducted by an autonomous unit, or by an independent firm, whose objective is to create a new product or a new technology and then develop it in a venture with the user firm. The innovation process is managed and owned by the independent research unit firm and financed by a financial partner, for example, a venture capitalist. CVC funds allow large firms to invest in start-up firms to gain a window on new technologies, stimulate development of complementary technologies, or encourage broader use of the investor's technology by establishing a *de facto* standard (Cohen 2000).
- M&As allow large firms to appropriate technology developed in small firms, even if they did not finance the R&D. While firms engage in M&As for many reasons other than accessing technology, the increasing numbers of small, R&D-intensive firms acquired by large high-technology firms indicates the growing importance of technology sourcing in such decisions. Firms can choose between developing a particular technology in-house or acquiring it from the open market through a merger or acquisition.

Both in-house R&D and external R&D organizational structures are observed in many industries. Moreover, the same firm may employ both organizational forms. For example, consider the pharmaceutical industry. A drug firm like Merck is investing mainly in in-house R&D although

[21] H Chesbrough (2000) Designing corporate ventures in the shadow of private venture capital. *California Management Review* 42(3), pp. 31–49.

some of its major rivals are outsourcing most of their research activities to biotechnology firms. Only about 5% of Merck's research spending ends up outside the firm's laboratories. For other top drug companies, however, the proportion of research done externally can reach 80%. Recently, some American pharmaceutical companies moved from in-house R&D to independent R&D by increasing their research joint venture agreements. These research joint ventures are contractual agreements for developing, producing, and selling a new medicine discovered by a biotech firm (Lerner and Merges, 1998).

Organizational Zoom . . .

Bang & Olufsen: The Struggle with Designers Free Reins

B&O, the high-end Danish producer of televisions and audio systems, while benefiting from a worldwide reputation for acclaimed design and luxury features, posted disappointed financial results for 2007 and the following years. On top of the economy turn down, the company faced a deep challenge with the arrival of the Apple's products and systems such as the iPod, iTunes, and iPhone, which drastically changed the rules of the game in the entire entertainment industry.[22]

Until 2009, designers, mostly external providers of advanced solutions to B&O, exercised almost complete authority over the new product development process. This long-established open-innovation posture allowed the products of the 80-year-old Scandinavian firm to hold places of honor in the collections of the world's greatest modern art museum. The company was in little danger of being imitated but showed very little ability to adapt with digital transformation, service system delivery, and interoperability of products. Adapting the development process to the imperative of the early 2010 and accelerating cycles proved difficult.

Already in 2004 a new unit, termed Idealab, was established to explore extensions to the firm's product development processes with a focus on computer technologies. But the acceptance of Idealab in the old company culture took longer than expected. So a series of organizational adjustments had to be undertaken in order to find the right balance between the early stage open innovation advantages and the contemporary necessity to fully integrate all phases of development.

[22]RD Austin and D Beyersdorfer (2007) Bang & Olufsen: Design-driven innovation. Harvard Business School case series No. 9-607-016.

The current trend of engaging in various forms of inter-firm co-operation allows businesses to nurture and benefit from the development of a wide range of new technologies without committing internal R&D resources to them. Outsourcing implies the transfer of R&D-to-R&D performers outside the firm, with a commensurate decline in internal R&D. Current modes of technology sourcing may result less in a reduction of in-house R&D than in a change in allocation of internal R&D funding. Companies may dedicate increasing R&D resources to those activities in which managers believe they have the greatest capability. Rather than weakening (or hollowing out) the R&D capabilities of large firms, external sourcing may increase the efficiency of business R&D and innovation systems overall by allowing a wider range of organizations to concentrate their R&D efforts in areas of relative strength.

Vertical vs. Horizontal Integration

Vertical integration is a corporate structure whereby product development through a supply chain is controlled or owned by one company, such as when a manufacturer owns its supplier and/or distributor. This structure allows the company to design and manufacture components, subsystems, and final product assembly before selling it to the customers. For example, GM or Ford's power train systems are vertically produced because the companies perform the aluminium casting of the engine blocks and pistons, through to the assembly, and integration into the vehicle.

Opposite to vertical integration, horizontal integration is a strategy where a company focuses on one aspect of the value chain and integrates all product offerings at that stage of the supply chain. A successful example is Dell Computers: The Company offers a full range of products but focuses on a thin band of the supply chain, while it continuously interacts with a dynamic supply chain. Vertically integrated companies normally have strong product development capabilities. For instance, Tetra Pak has 60 years of experience leading the industry in cardboard containerization and DuPont has a core expertise in commercializing chemistry breakthroughs. Traditionally, these companies have been working in a closed innovation fashion, based on strong internal capabilities, but due to market forces they are now embracing open innovation.[23]

[23]http://www.innovationmanagement.se/2014/04/04/organizational-processes-and-structures-supporting-open-innovation/.

Organizational Zoom ...

Nottingham Spirk: Vertical Innovation

Nottingham Spirk, an open innovation and product development firm, coined the phrase "vertical innovation" in 2011, to describe the proprietary process developed over the 40 years at the firm.[24] Vertical innovation is both a process to encourage efficient problem solving and a corporate culture, where all development activities are handled by one cohesive and experienced team of experts.[25] Nottingham Spirk serves clients as an open innovation partner and uses a vertically integrated model to execute on development, hence a 'vertical innovation' strategy.

This group leverages the skills and talents of many individuals to quickly move product development forward. This process focuses on continuous improvement through each phase of a development program. Their practices begin with understanding target markets and customers, and cover subsequent processes from research to product design to prototyping to testing, which are all conducted at their innovation center to ensure efficiency, confidentiality, and intellectual property protection.[26]

Open innovation is about leveraging collaboration between companies that are ready to contribute to a common project. Horizontally integrated companies can utilize their market position and network of suppliers/clients to become the nexus of the evolving open innovation network. Suppliers and small actors in the innovation network will submit ideas to the company. In this way, central companies like Dell can greatly increase the number of ideas coming into their product development funnel, while they can simultaneously decrease their R&D costs. They can utilize their value chain position to attract the type of ideas they want and commercially leverage the best ideas by integrating them into their new products.

Centralized vs. Decentralized Organization

Both centralized and decentralized ways of organizing open innovation have their pros and cons. In most cases, companies have a central organization of open innovation at the beginning. It then becomes more like a hybrid

[24]http://trademarks.justia.com/855/51/vertical-innovation-85551798.html.
[25]http://www.nottinghamspirk.com/vertical-innovation/.
[26]http://www.nottinghamspirk.com/.

structure once companies understand that they need to impact more people in order to bring openness into the firm's innovation DNA. A lot of the companies move toward hybrid models as their organizations have matured at open innovation.

The centralized organization is supposedly the easiest to manage, but it is sometimes difficult for the business units to feel enough ownership. The author concludes that having some form of hybrid is probably the best, as it provides ownership of the businesses, yet it gives a center of expertise. If business buy-in is not an issue, it is more efficient to have a centralized function, since in this way an open innovation expertise is used optimally with no need for duplication across the organization.

Formal vs. Informal Organization

A fertile organizational structure and a collaborative culture are essential to the successful implementation of open innovation. Formal organization of open innovation does not necessarily imply a hierarchical structure. Companies can have formal structures that are not hierarchical and extremely bureaucratic (e.g., matrix type structures). There is no evidence, which formal types of organization fit best as archetypes for open innovation, but it is obvious that we should think in a nuanced fashion about the distinction between formal and informal structures. Yet, many formal structures are just not workable in practice for implementing open innovation.

Instead of formal structure and processes, utilizing the informal personal networks could be a good start to get open innovation implemented. One member referred to one company that has a formal structure and a parallel informal structure and managers move between the two, following the first for compliance/reporting and using the latter for commitment/results. With such 'hybrid structures' combining formal and informal and/or centralized structure, companies can get open innovation implemented successfully in organizations.

In this section, we are not trying to compare the effectiveness of the use of open innovation in different organizational structures. We believe that successful open innovation has to be organized differently according to the existing structures and capabilities. For example, Ihl, Piller and Wagner (2012) assert that higher performance gains from open innovation by aligning internal organizational structures in terms of lower specialization as well as higher formalization and decentralization. Based on a sample of German manufacturing firms, they found that for firms with high internal R&D intensity, lower specialization is an effective mean to turn a substitutional effect of open innovation into a complementary one, while for firms with low internal R&D intensity; higher formalization and decentralization are an effective means to enhance the substitutional effect of open innovation.[27]

Implementing open innovation successfully also requires organizational innovation: New structures are required for open innovation to thrive, rather than simply survive. A successful example is Danone's communities of practice: It is an internal 'online' platform where employees could exchange ideas, answer each others' questions, etc. on a voluntary basis. Engagement rates were very high. Siemens is moving in the same direction. Hence, the 'semi-formal' can take a life of its own, and might eventually become formal, which again, may not be the best thing.

DEVELOPING ORGANIZATIONS FOR SERVICE MIX ACTIVITIES

Some functional separation is necessary for developing the professionalism and the quality of the work carried out whatever unit is considered. On the other hand, integration is the condition for ensuring coherence in managing the product throughout its lifetime, as well as for controlling the total cost of its development and its consumption. So attention has to be focused on the identification, the categorization and the establishment of integrative units in any Service organization. The concepts of integration and separation, permanently linked together, are richer than the pair centralization/decentralization, even though the latter forms the basis for delegation of functional responsibilities within the Service organization.

Co-ordination is required in virtually every activity that comprises the service function. Thus, as an example, modifying a replacement part should lead, theoretically, to adapting the documentation identifying it, to modifying the kits and updating stock, to transmitting to the replacement logistics function the anticipated rate of availability, as well as to managing the circulation of these parts and rebuilding the former models to replace them, etc. For most of the after-sales managers whom we have interviewed in a general survey, such serially-connected actions prove to be very

[27] C Ihl, F Piller and P Wagner (2012) Organizing for open innovation: Aligning internal structure and external knowledge sourcing. *Paper presented at the DRUID 2012*. CBS, Copenhagen, June 19th.

difficult to carry out because of the absence, within their organizations, of integrative units explicitly responsible for guaranteeing this sequence of activities.

After some empirical analysis carried out in several companies, we were able to identify a series of elementary tasks that could constitute the mission for an integrative organizational unit undertaking Service Mix activities. There are, of course, many ways to group these elements within such a unit. Our proposal is based on a view of Service Mix activities as a value chain that accompanies the concepts of before-sale, during-sale, and after-sale time periods. Activities can be grouped into four clusters:

- Design is the firms answer (on paper) to the needs of its customers over time, i.e., the conceptualization and planning of all tangible products and components as well as all support services that accompany those tangible items throughout the product's lifetime. This, of course, goes well beyond the usual organizational definition of design function, and includes the structuring of the process by which customer needs, over time, become known to the firm, along with the design of financing options over the lifetime of the product, and the planning of rebuilding, remanufacturing, or retrofitting activities, if necessary later in the product's lifetime. These design activities exist, not only to effectively serve customers' needs, but also to create efficiencies and synergies internal to the firm.
- Preparation for use brings together all activities that allow the customer to obtain the physical product, together with all services that accompany it, at the customer's premises. These activities include initial manufacture, physical distribution, the provision of compatible accessories if appropriate, and the installation of equipment in the workplace. Again, this goes beyond the activities' traditional definition. As a small but illustrative example in consumer electronics industries, some firms have deemed it fitting to provide batteries with their products rather than expect the consumer to make a separate purchase of those necessary accessories.
- Information activities encompass all those that, on one hand, gather and encapsulate knowledge of customers' needs and how they use the product and, on the other hand, transmit data describing the characteristics of a product or directions for its proper and most effective usage. Examples of the latter category range from customer training for complex equipment, to the printed instructions as to how to set a new watch or to the recipe

on the back of a box of cake mix. However, these are relatively mundane examples. More creative, for example, is the establishment by L'Oreal, of an information service available over phone during working hours, for hairdressers to call when they encounter problems with L'Oreal products, e.g., reactions producing unexpected color.

- Upkeep is the set of activities more typically associated with a company's Service Department and includes maintenance, reconditioning, and equipment loans, etc. Years ago, for example, in the difficult auto supply industry, the Chausson group developed a business reconditioning used car radiators that has been profitable, while allowing the company to establish better relationships with automobile dealers.

Our attempt to implement this concept within High Tech Equipment companies takes into account those tasks to be performed within the Service Department that contribute to each of these four categories of activities. This is a matrix structure of sorts, with each individual responsible for both a group of products and one category of activities. The need for strong cross-functional links should be clear. Personnel concerned with the planning functions which we have grouped under Design must cultivate strong ties with marketing, R&D, and Finance. Those concerned with Preparation need to interact frequently with Manufacturing and Physical Distribution. Those concerned with Information must not only deal directly with customers, but also with sales, in addition to other functional groups. Those concerned with Upkeep, must in point of fact, deal with virtually every internal function. Correct maintenance, on the customer's premises, requires understanding usage needs (Sales), manufacturing methods, and recent engineering changes, etc. In our view, this of course, leads to a great increase in the complexity of co-ordination which is necessary for success in this competitive marketplace.

In general terms, decisions regarding the evolution of any organizational structure must contribute to increasing a company's abilities to constantly innovate. However, innovations are hard to plan; often they result from a combination of luck and of necessity, a necessity that is felt by consumers. Therefore given the current and future challenges that companies have to meet, the capacity of the organizational structure to foster innovation is becoming critical. Today, leadership's role in creating growth is all about "design", i.e., the design of the organizational structures that lead to agile, flexible cultures which support lateral collaboration and rapid innovation cycle. The challenge involves knowing what kinds of organizational designs will best support your business models and strategies, and how to instill these as sustainable capabilities.[28]

[28]Prof. Roy D. Shapiro, Ziying Huang, and Marwyn O'Keeffe deserve much credit for their assistance in the writing of this chapter.

Chapter 7

MANAGING TALENT FOR INNOVATION

Lying at the heart of innovation, and pivotal to its creation, are the people who *generate* innovation in the organization. Innovation is the product of "knowledge workers", a term coined by Peter Drucker and one that highlights the monumental transition of our contemporary economy from one that derived value principally from manual labor, to one that is based largely on the knowledge and services output of workers and the "ideas" they produce which result in the creation of economic value. Innovation is essentially a people-centric mechanism and today the greatest challenge for organizations and societies lies in the ability to harness the creativity of employees and stimulate their motivation to produce ideas of value. This marks a distinct departure from the manufacturing dominant economic model of the industrial age. The prerequisite of innovation is either the generation of new knowledge or, as is more often the case, the combination of existing knowledge in new "entrepreneurial" ways as originally explained by J.A. Schumpeter and then by P. Drucker.[1]

> "*Capitalism is undergoing an epochal transformation from a mass production system where the principal source of value was human labor to a new era of 'innovation-mediated production' where the principal*

[1]JA Schumpeter 1934 (2008) *The Theory of Economic Development: An Inquiry into Profits, Capital, Credit, Interest and the Business Cycle.* Translated from the German by Redvers Opie, New Brunswick, USA and London, UK: Transaction Publishers; P Drucker (1985) *Innovation & Entrepreneurship: Practices and Principles.* New York: Harper & Row.

> *component of value creation, productivity and economic growth is knowledge".*
>
> — Richard Florida & Martine Kenney[2]

Moreover, the generation of innovation normally involves a group of people and rarely takes place in the hands of a single person. On the other hand, invention can often be the work of one person and historically one person has often been responsible for an invention. But inventions do not always result in the creation of market value. For an idea to be created, developed, and transformed into the creation of value for the market, the result is quite often the fusion of exchange of multiple skills and knowledge, and collaboration among departments. As A. G. Lafley, the Chief Executive Officer of Procter & Gamble (P&G), claims, "*You need creativity and invention, but until you can connect that creativity to the customer in the form of a product or a service that meaningfully changes their lives, I would argue you don't yet have innovation*".

We know that some of the best ideas for innovation come from employees themselves. Amazon employee Charlie Ward, a software engineer, came up with the idea for Amazon Prime- the company's all-you-can-ship loyalty program which guarantees delivery of products within two days for an annual fee of $79. The idea was then collaboratively developed by a team of employees and closely followed by Jeff Bezos, the CEO. This game-changing service innovation is thought to be responsible for the company's share prices rising 296% in two years since the launch and the main reason behind Amazon's sales rise by 30% during the recession.[3]

So the important question is how do we manage and motivate "knowledge workers"? How can companies manage the intangible, invisible cognitions of the mind? And how can the generation of ideas and their transformation into innovation are handled successfully? By instilling an appropriately balanced chemistry, and by mobilizing creative thinking, expertise, and motivation, organizations may be able to innovate.

---

[2]R Florida and M Kenney (1991) The New Age of Capitalism. *Futures*. In MPRA (2009) *The Knowledge Economy*, MPRA Paper No. 18256, October 30th.
[3]BusinessWeek (2010) What's in Amazon's box? Instant gratification. November 24th.

LEVERAGING THE FIVE DRIVERS OF THE INNOVATOR'S MOTIVATION

The motivation of people to innovate is of critical importance in the context of the changing competitive environment that is governed by the ability of an economy to generate value from knowledge and services, as opposed to physical labor and manufacturing. As Peter Drucker surmised so aptly, "*continuous innovation has to be built into the knowledge worker's job*".[4] The challenge of motivating the human mind to produce 'ideas' is decidedly more complex than motivating people to work faster or more efficiently on a physical product. Indeed, as A.D. Amar notes, "*since productive knowledge in organizations is ultimately a product of the human mind, it cannot be manipulated like sophisticated machines, fancy systems, or efficient controls. Organizations that desire to use knowledge in their products, processes, and services, have to know how to engage the human mind in their operations*".[5]

> "*Imagination and commitment are things that people choose to bring to work every day — or not*".
>
> — Gary Hamel

Today, motivation and management are indeed more complex than the Taylorist approach of the industrial age which was categorized by repetitive division of labor techniques, incentives tied to production, and a system that enabled workers to make as few mistakes as possible. As the author of *The Creative Class,* Richard Florida, notes, "*Creative work cannot be taylorized like rote work in the old factory or office*".[6] Florida offers several reasons for this. Creative work is not repetitive and therefore does not lend itself to "scientific management". Secondly, you cannot "taylorize" what you cannot see; the work of the knowledge workers is intimately linked to the intellect of the individual. Furthermore, the control, systemization and the sense that one is simply a "cog in the machine" has tended to incite rebellion amongst workers. Many observers, researchers and managers argue that

[4]PF Drucker (1999) Knowledge-worker productivity: The biggest challenge. *California Management Review* 41(2), pp. 79–94.

[5]A.D. Amar (2004) Motivating knowledge workers to innovate: a model integrating motivation dynamics and antecedents. *European Journal of Innovation Management.* 7(2), pp. 89–101.

[6]RL Florida (2002) *The Rise of the Creative Class: And how it's Transforming Work, Leisure, Community and Everyday Life.* New York: Basic Books.

coercive or overly controlling management tends to be counter-productive; people respond negatively when they feel that they have been pushed into a certain type of behavior. Instead, employees need to feel a certain level of choice, autonomy, and flexibility at work, to be self-determining. Self-determination occurs when the employee experiences choice in his actions at work; he does not feel overly controlled by the organization and has the freedom to realize objectives in an autonomous way. Self-determination is supported by the participation of the employee in decision-making, allowing freedom for individual initiatives, and transparent communications. Threats of punishment, surveillance, and evaluations tend to be experienced as controlling, and thus negatively affect the employee's sense of choice, self-esteem, intrinsic motivation, and perceived competence.[7]

> *"How do you build organizations that are as nimble as change itself? How do you mobilize and monetize the imagination of every employee, every day? How do you create organizations that are highly engaging places to work in?"*
>
> — Gary Hamel

Today productivity and the meaning of work have a different, less tangible quality where the "products" of a knowledge economy do not lend themselves to metrics or a carrot or stick approach toward motivation. The effective management of motivation is inextricably linked to meaning, purpose, and value, to inspiring people to innovate, to solve problems, to create, and to become collectively and collaboratively engaged. Specifically, we identify five aspects that companies should focus on to motivate their employees to innovate.[8] Figure 7.1 shows these aspects.

Acquire

People want to acquire: Knowledge, success, prestige, money and so on. In the case of knowledge workers, financial incentives and bonuses may not be that important. In fact, Peter Drucker, who studied the knowledge worker in great depth, also argued that financial incentives, orders or negative reinforcements do not motivate knowledge workers. Instead, people

[7]EL Deci *et al.* (1989) Self-determination in a work organization. *Journal of Applied Psychology* 74(4), pp. 580–590.

[8]The construction of the suggested five-stage framework for innovation motivation has benefited from inspiring discussions with Pascal Minella, founder and chief executive officer of furious intents, change & innovation management consultants in Paris.

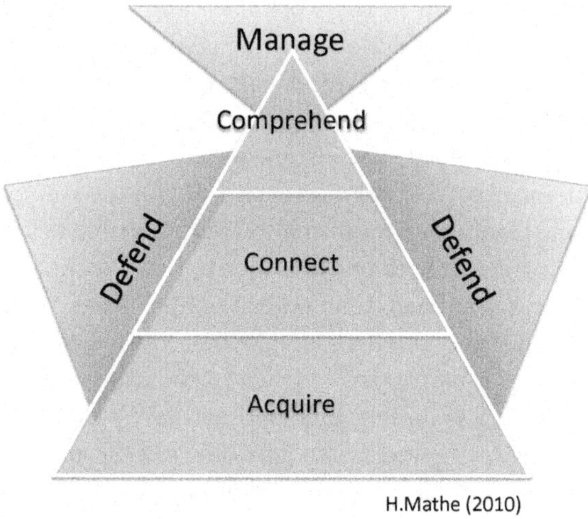

Figure 7.1: The Five Motivators of the Innovator's Motivation.

who get the chance to work on innovation may be more driven by the passion of what they are doing. They like intellectual challenges and being able to work with other qualified and competent people in order to reach their goals.

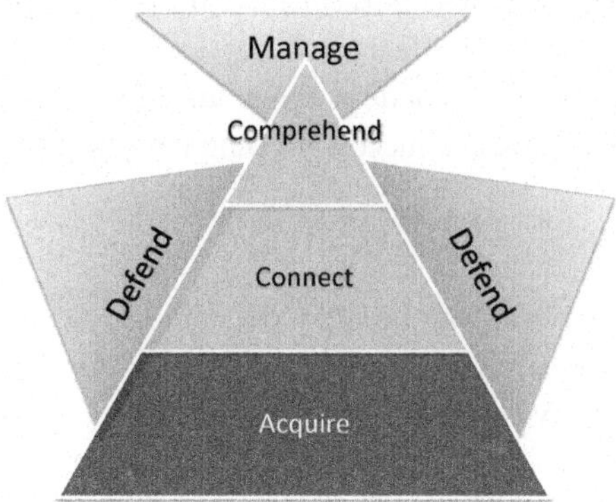

Goal-setting, a management method developed in the 1980s by Edwin Locke and Gary Latham, is an important tool for motivation that enables employees to structure and prioritize their work, and motivates them to complete tasks and improve performance. Goals, defined as desired end states or aspirations for the future, play a causal role in employee action and thus, according to Locke, represent one of the most powerful determinants of motivation in the workplace and conscious goals are thus expected to positively affect action. Stretch goals, a practice that originates in the Japanese philosophy of *Kaizen*, or continuous improvement, made famous by its successful evangelization in the Toyota lean manufacturing model, can be an effective way of encouraging employees to strive for greater performance or to develop creative or innovative solutions within the organization. Goals and performance reviews represent an important method of motivating superior performance in the context of the knowledge economy. Google, a company that enjoys a high level of motivation and discretionary innovation amongst employees, recognizes the importance of agreeing goals with employees and complementing this process with a frequent feedback loop including four performance reviews per year.

To fulfill innovators' need to acquire knowledge and to achieve their goal of creating innovations; companies should also provide their employees enough autonomy and encourage the freedom to innovate or to pursue an activity that interests them. Autonomy and self-determination can have an impact on the generation of innovation by individual employees. Google have identified this phenomenon amongst its engineers. They tracked the progress of ideas that they had backed versus ideas that had been executed in the ranks without support from above, and discovered a higher success rate in those that had been developed without the support of upper management. Similarly, it was noted that Philip Rosedale, the founder and chairman of Linden Lab, the fast-growing company that manages Second Life, states that his workers benefit from an enormous level of autonomy, and he finds that the greatest successes come from workers' own initiatives.[9]

Connect

People by nature seek to forge bonds with others near them to feel secured, and exchange information. Today, by necessity, more people build those bonds with non-relatives; people spend more time at work and find these "psychological kins" in the workplaces.[10]

Connections with others are of great importance to innovators. Undoubtedly, innovation is always the product of teams. By exchanging information and brainstorming in teams, people can clarify their ideas

[9]TM Amabile and M Khaire (2008) Creativity and the role of the leader. *Harvard Business Review* 86(10), pp. 100–109.

[10]Wall Street Journal (2002). Along with benefits and pay, employees seek friends on the Job. February 20th.

and accelerate their thinking. Companies should create an environment that encourages their employees to use their connections and networks to mobilize resources and build alliances, both internally and externally.

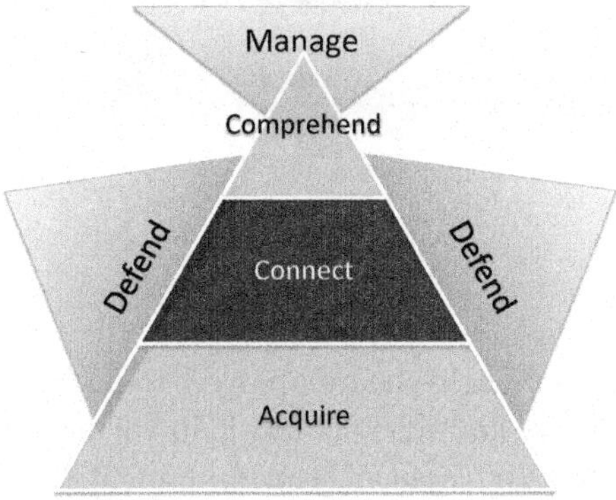

Without the sharing of knowledge inside or outside the organization, an individual relies only on his or her own existing knowledge for innovation. If the culture of the organization does not create the right incentives for sharing, then knowledge can become isolated and will not benefit from 'cross-pollination'. Optimizing the flow of competencies within the organization is central to the management of knowledge and the generation of innovation.[11] Frank Mulhern of Northwestern University underlines this perspective stating that organizations succeed because information and actions flow from person to person in what he calls the *"flow of performance"*. This implies that the collective knowledge generated by the flow of information and knowledge sharing between the people in the organization may be of greater value than individual talent.

Comprehend

As the paradigm of innovation is migrating from a traditional framework of pure product innovation to a more complex model combining service and product, innovators need to see with a fresh perspective how innovation evolves and works. Every innovator would like to be an iconoclast, but the difficult issue is "how?" — how to be an iconoclast?

---

[11]DM Amidon (1997) *Innovation Strategy for the Knowledge Economy: The Ken Awakening.* Waltham, MA: Butterworth–Heinemann.

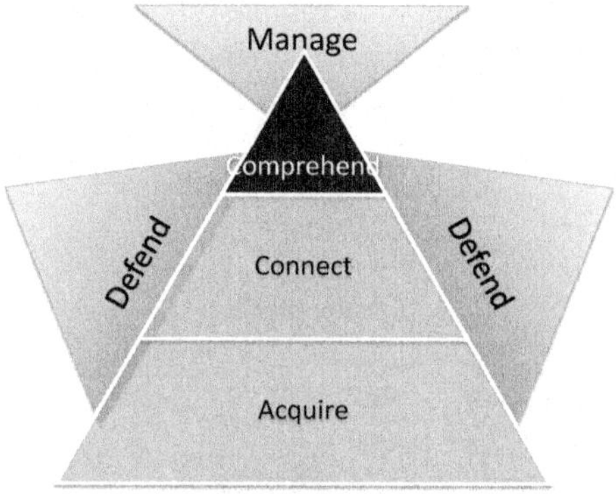

Organizations should help their employees to understand the company's strategies and missions, and also should help to master relevant methodologies and techniques. Without clear communication of corporate visions and an innovation operating model, innovation teams could lack direction and proper support from the executives who do not really understand how to approach the modern kind of innovation as we see it today. In addition, with the primary focus shifting from the product itself to a "utility in use" notion, new methods and approaches are required for innovators to understand customers' needs better. For example, companies could provide some training sessions on design thinking for engineers to comprehend the questions of why and how to conduct user-centric innovations.

Organizational Zoom . . .

Alessi: Implementing a Design Thinking Framework

"It doesn't matter whether you're selling teddy bears or aircraft engines, your company could benefit from a deep intuition for customers that transcends explicit data".

Alessi, an Italian accessories designer that produces iconic designs, has a strong tradition of artisanship and a unique design-driven innovation culture in which its designers are acutely conscious of their freedom to create as they see fit. The design process at Alessi aims to be breakthrough rather than simply responding to customer demands directly.

As Alberto Alessi describes the approach to design at Alessi "*Our way is closer to the way of Pablo Picasso. Imagine Picasso waking up in the 1920s on a nice, sunny morning in a village on the Côte d'Azur and feeling strongly the wish — the need — to start painting. So he starts painting. But he's not asking himself, "To what target customer will I address my new painting?" Picasso shows us a completely different approach: starting from yourself, as a creator, and using your sensibility and your intuition in order to touch other people's hearts or sensibility or intuition*". Indeed, it was Pablo Picasso who said "*I am always doing that which I cannot do, in order that I may learn how to do it*".

Truly innovative design originates with the designer's creative insight and at Alessi the creation process starts from intuition, not from market research. While the process can benefit from a deep understanding of the user, it goes beyond the straightforward dialectics of developing products corresponding exactly to the desire of the customer. Roberto Verganti, the author of Design-Driven Innovation asks this question: "*How to create innovations that customers do not expect, but that they eventually love*?"[12]

Defend

People also need to feel identified with the organization, believe in its purpose and functions, and feel part of the company mission. Drucker argues that one of the most effective ways to motivate knowledge workers is to treat them as "*de facto* volunteers", the idea that workers are committed to the company and interested in its future because they want to be there, and not simply because they are paid to be there. This approach is strongly linked to giving employees the respect they deserve, involving them in the strategic decision-making processes of the organization, and instilling people with a sense of meaning and inclusion. The organization is the sum of its parts and is not run by one individual. As Drucker states, "*what motivates knowledge workers is what motivates volunteers. Volunteers, we know, have to get more satisfaction from their work than paid employees, precisely because they do not get a paycheck*".

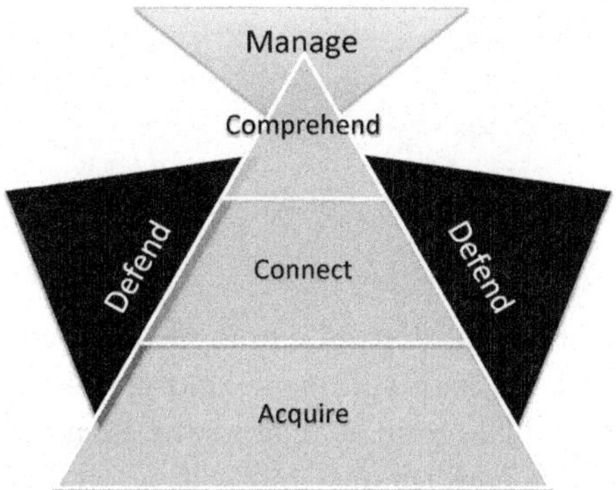

[12]http://www.designdriveninnovation.com/.

Industry leaders reputed for their positive working environments like Google in IT and Southwest Airlines in air travel have learned that creating a sense of real meaning for employees in their jobs is integral to fostering commitment and a desire to generate innovation. When people feel like an important part of the "big picture" for the organization, they realize the importance of their contribution to innovation and feel motivated to participate in this process. Innovation is not something that should be confined to the boardroom or to a "select" group of employees; people across the organization should feel welcome to contribute ideas that may lead to innovation. Giving employees this opportunity sends the message that the company values the individual and their ideas.

Manage

We also need to be managed by adequate leaders and supporters, people with the right behavioral traits: People who combine a mix of emotion and realism; managers who accept the taking of risks and failures, the factors that are inherent to innovation; individuals who have the courage and the integrity to stop projects even when large investments have been made; and people who have a talent for building and steering teams.

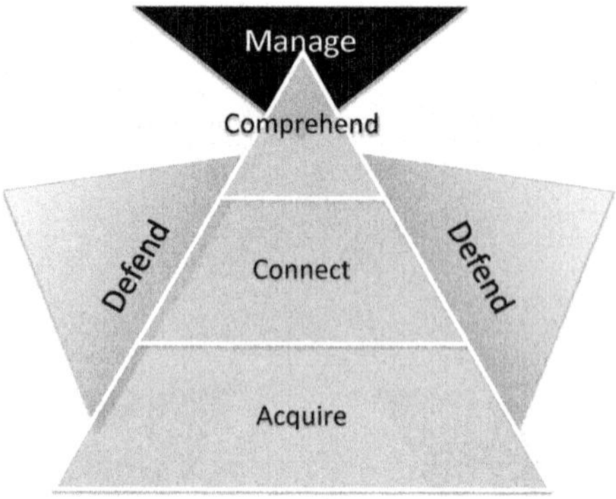

With the understanding that an employee's relationship with the boss plays an integral role in motivation, or more specifically, demotivation (Herzberg), the value of a good manager was identified to be of central importance to employee performance, motivation, well-being, and retention.

For years Google followed an intellectual-driven approach for hiring good managers: Brilliant engineers were sought and the brightest of them were promoted into leadership positions. The "discovery" has since been made that the most intelligent engineers do not always become the most effective managers, and the difference in employee satisfaction was clearly observable through the quarterly reviews in which there were huge variations in the ratings that employees gave to their bosses.[13]

Organizational Zoom...

Google's Project Oxygen: How Does Google Select its Managers?

Laszlo Bock, Google's SVP for Human Resources, surmises "*our best managers have teams that perform better, are retained better, are happier — they do everything better. So the biggest controllable factor that we could see was the quality of the manager, and how they sort of made things happen. The question we then asked was: What if every manager was that good?*"[14]

The observation led Google to implement an innovative management study entitled "Project Oxygen" to uncover the characteristics deemed essential for effectively managing "Googlers" in the unique Google culture and operations. One striking conclusion is the illumination that technical knowledge in fact may not be the top priority as it was always deemed, i.e., a manager does not necessarily need to have a greater technological knowledge than his or her subordinates. As Lazlo Bock confirms, "*we'd always believed that to be a manager, particularly on the engineering side, you need to be as deep or deeper a technical expert than the people who work for you. It turns out that that's absolutely the least important thing. It's important, but pales in comparison. Much more important is just making that connection and being accessible*". Google conducted this analysis based on internal data that makes it uniquely relevant to the company.

Google has always had a unique selection process for new recruits, but many of the qualities previously considered essential have slipped to the bottom of the list, such as the technological brilliance. It nevertheless features on the list.

[13] M Bryant (2011) Google's quest to build a better boss. *New York Times*, March 12th.
[14] *ibid.*

Google's Project Oxygen: the 8 most important skills of a Google manager
Be a good coach
Empower your team and don't micromanage
Express interest in your team members success and well-being
Be productive and results-oriented
Be a good communicator and listen to your team
Help your employees with career development
Have a clear vision and strategy for the team
Have technical skills so you can advise the team

DEVELOPING A CULTURE OF INNOVATION

Henry Mintzberg once wrote that "*a company without a compelling culture is like a person without a personality — flesh and bones but no life force, no soul*".[15] Culture is composed of the atmosphere, the unspoken rules and values, and the experiences that the people in the company have lived through together.[16] As Cha and Chatman (2002) note, "*strong culture enhances performance in two ways. First, it energizes employees by appealing to their higher ideals and values, and by rallying them around a set of meaningful, unified goals. Second, it boosts performance by shaping and coordinating employee behavior*". Indeed, a compelling culture not only creates a working environment where people actually want to be, but it creates the right environs for innovation and creation. Google is a classic example of a company that has re-written the rules of corporate culture by implementing an unconventional, playground-like environment and culture that inspires people to feel comfortable and bring out their creativity. In Google, the start-up culture is nurtured through examples set from the top–down. Patrick Pichette, Google's Chief Financial Officer, flies coach class, and responds directly to e-mails from fellow "Googlers" every day. As far as he is concerned, this "*crushes the idea of bureaucracy*".[17]

[15]H Minztberg (2009) Rebuilding companies as communities. *Harvard Business Review*, July–August.

[16]S Cha and J Chatman (2002) Mastering leadership: Culture of growth. *Financial Times*, November.

[17]Manyika (2011) Google's CFO on growth, capital structure, and leadership. *McKinsey Quarterly*, August.

He spurns the notion of business units because they put people in autonomous "ownership" positions that hamper creative flexibility.

Edward Schein, an MIT professor who extensively studied the subject of organizational culture, argues that, as a historically defined construct, culture can prove to be very difficult to change.[18] Instilling a new energy and collective ideal for a company that has never before benefited from the communal motivation of a constructive organizational culture can thus be challenging. When Carlos Ghosn, CEO of Renault and nicknamed "le cost killer",[19] took over as CEO of Nissan, the company had been suffering from collective dispirit, he tried to rewrite the *status quo* in the company and kick-start a performance driven energy that would guide behavior and overturn the values of the company. He said that an extensive "emotional mobilization" of the members of the company was necessary. The first thing that he changed was the language that the employees spoke in work which was clearly a colossal challenge for the Japanese speaking employees. The goal was to demonstrate to the people that the company did not have stronger affiliations to the French or Japanese side of the company; it was a global company. It was the first step in a "social movement" to restore people's pride in the company and reignite the culture of innovation.[20]

Specifically, an innovation-supportive culture should foster teamwork and collaboration, support risk-taking, encourage initiative-taking, and recognize and foster individual creativity.

Fostering a Culture that Supports Collaboration

There is a general understanding amongst business leaders that innovation rarely results from individual "geniuses" experiencing moments of "Eureka".[21] Breakthrough innovation ideas are more often the product of collaborative ingenuity. The importance of collaboration within the company is widely known by firms seeking to innovate. In the most innovative organizations, collaborative idea-shaping takes place in many situations, both formal and informal: In brainstorming sessions; through

---

[18]EH Schein (2005) *Organizational Culture and Leadership, 3rd Ed.* Hoboken, NJ: Jossey-Bass, Wiley.
[19]D Fonda (2003) Motor trends: Le cost killer. *Time,* May 19th.
[20]R Hayagreeva (2009) Market rebels and radical innovation. *McKinsey Quarterly,* January.
[21]JR Ryan (2010) Leadership Skills for Driving Innovation. *BusinessWeek.com,* 6.

knowledge sharing web-based tools; by bouncing ideas of fellow colleagues, customers or even random strangers; discussions in the stairwells and the staff restaurant. Indeed, as Lawlor (2000) notes, "*The innovative culture thrives on the free exchange of ideas*".[22]

Organizational Zoom ...

Pixar: Breaking Down Barriers and Individual Fears

Pixar's CEO Brad Bird underlines the importance and the challenge of breaking down barriers and the individual fears among employees so that they feel comfortable collaborating with one another, and more importantly, challenging one another's ideas.

For the movie *The Iron Giant*, Bird inherited a team motivation that was at rock bottom following a failure with a previous project. The way Brad Bird succeeded to motivate this group of people was to step down from his position to be on an equal footing with the team, and take weekly sessions with the group where each employee's work and contribution was analyzed in front of the whole group. Contributions were highly valued and he also encouraged people to speak up if they had additional ideas or if they disagreed with certain procedures. However, he admits that "*it took two months for people to feel safe enough to speak up*".[23]

Creating a collaborative environment is central to motivating people to innovate effectively and to perform well within the organization. Jack Welch famously coined the phrase "boundary-less organization" referring to a more appropriate firm organization that would be trans-disciplinary in nature with greater interactivity across projects and functions, cultures, and geographies. This "boundary-less organization" is arguably the format that is most relevant today for generating innovation in the firms and driving the cross-pollination of ideas between people and departments.

Henry Minztberg, too, has asserted the need for companies to foster a sense of "*communityship*" within the organization, for companies to rebuild themselves into places of engagement, where people are committed to one another and their organization. Minztberg advocates a move away from the individualist behavior and toward practices that promote trust, engagement, and spontaneous collaboration aimed at sustainability.

[22]Deloitte Touche Tohmatsu (2005) Fostering an innovative culture, sustaining competitive advantage. http://www.cecpartners.com/articles/CECP%20Strategy/GES_FosteringInnovativeCulture.pdf.

[23]R Hayagreeva, R Sutton and AP Webb (2008) Innovation lessons from Pixar: An interview with Oscar-winning director Brad Bird. *McKinsey Quarterly*.

According to Minztberg, fostering this sense of "*communityship*" means "*caring about our work, our colleagues, and our place in the world*", and in turn being inspired by this caring.[24]

Similarly, the CEO of Wells Fargo highlights the importance of improving team cohesiveness across the company through the approach entitled "One Well's Fargo". The company is seeking to improve the performance and health of the organization through a process known as "mining the seams" of the organization which is nothing but working together more effectively across the company's lines of business to break down "silo thinking" and to improve the customer experience and fulfill more of their financial needs.

In addition, as opposed to a closed and siloed culture, teamwork and collaborations between employees and departments can foster organizational learning through sharing and passing on of knowledge, and hence increase the innovative capability of a company. According to Herbert Simon, an organization learns in only two ways: (1) By the learning of its members, or (2) By bringing in new members who have the knowledge which the organization did not previously had. He qualifies the statement with a crucial component impacting the efficacy of organizational learning. The relationship between the tacit knowledge of one individual with that of other individuals within the organization may have a great bearing on how the organization collectively learns and makes use of this knowledge. In other words, the way in which people interact, collaborate, and share expertise has an impact on the capitalization of such knowledge. The culture of the organization plays a role in defining the codes of knowledge sharing: What an individual actually learns in an organization is very much dependent on what is already known to, or believed by, other members of the organization and the kinds of information that are present in the organizational environment.

Organizational Zoom ...

BlaBlaCar: Creating a Community for Carpooling

While gas prices in France recorded their historic high with an average of 1.54 Euro per liter of super unleaded petrol, Covoiturage.fr — the most popular carpooling website in France, celebrated its one million

---

[24]H Minztberg (2009) Rebuilding companies as communities. *Harvard Business Review*, July–August.

membership base in April 2011. From 2004–2012, the company claims to have allowed its users save Euros 182 million and induce more than 30 million sharing experiences. Renamed BlaBlaCar in 2013 to facilitate its international development, the company served 10 million users in ten countries for 2014 and took the European leadership in carpooling. In July 2014, BlaBlaCar raised US$ 100 million from Index Ventures with the objective to rapidly become the worldwide leader.

At the young startup, people, their wishes, needs, and general well-being, are a core value. Employees are regarded as valuable individuals with strengths and weaknesses making them unique. Customers also are carefully listened to and the company aims to develop a community where people interact and share resources in order to save money but also to reduce the negative impact of mobility of the environment. They not only integrate events such as festivals, sports and others in their agenda on the site, but also offer the possibility to create private events, like wedding or birthday parties, so that users can organize their travel.[25]

Frederic Mazzella, BlaBlaCar chief executive who took over from Vincent Caron in 2006, is a former NASA scientist and Stanford graduate. He quickly realized the importance of networking as a key ingredient for building a successful startup. As an avid physicist, perhaps he views the world connect in a different way than a classical management graduate, or perhaps his personal conviction for networking stems from his personal experience as an entrepreneur, a quality that he honed in an INSEAD MBA.

Whatever the case, networking has been a key part of his strategy in the development of his car-sharing community. In the company's offices in Paris, France, he has set up a hub for startups and each Friday everyone gets together for breakfast to listen to one of the startups present their company. Why? Bringing together creative people, most of who are energized by a similar motivation to create and to succeed, is mutually beneficial. They can share ideas, contacts, and stimulate the production of new ideas and creative solutions. Moreover, growing a startup can be tough and living this experience with the support of others in the same situation can be infinitely motivating.

A key aspect of organizational learning is 'internal learning', that is, the transmission of information from one organizational member to another. Simon (1991) argues that individual learning in the organization is not a solitary phenomenon, but actually a social occurrence.[26] Alfred Bandura referred to this key determinant of individual behavior

[25]H Mathe, F Mazzella and H Schneider (2012) Innovation at Covoiturage.fr, creating a community for carpooling. *Innovation & Society* 16, February.
[26]HA Simon (1991) Bounded rationality and organizational learning. *Organisation Science*. Vol. 2, No. 1. pp. 125–134.

as Social Learning: We learn from cues in our environment that are set by the culture and behavior of the people in the organization. Bandura emphasized the underlying influence of learning in the organization: "*If knowledge and skills could be acquired only through direct experience, the process of human development would be greatly retarded, not to mention exceedingly tedious, costly, and hazardous. Fortunately, people can expand their knowledge and skills on the basis of information conveyed by modeling influences*".[27] Modeling influences refer to the codes and examples set by the organization and can be leveraged to guide the practices of knowledge sharing across the organization. IT platforms also play a role but if the tacit network of collaboration is closed and opposed to cross-functional sharing of expertise then the fundamental inclination to share will not exist.

While hoarding or departmentalizing information can lead to a lack of trust and collaboration in the company, cross-disciplinary teams meshing designers, engineers, and different product development fields can open the innovation process to the possibility of "cross-pollination" of ideas. One entrepreneur referred to this phenomenon as the "Medici effect" in reference to the explosion in creativity in Florence when the Medici family connected people from different disciplines, namely scientists, poets, artists, architects, philosophers, and so on. The cross-pollination of ideas generated by the fusion of such a diverse range of disciplines caused new ideas and creativity to flourish which contributed to the birth of the Renaissance and stimulating invention.[28] Similar in the contemporary context, collaborative knowledge sharing links the learning and knowledge processes to enhance organizational learning and innovation. For example, P&G regularly bring together people from different disciplines to brainstorm over ideas; both internally in the company and externally through the company's open innovation platform. A team from pet food might well be able to solve a problem that has arisen in hair care products. Cross-disciplinary research is a way for companies to benefit from expertise within the company, to capitalize on knowledge that already exists, and to curtail risk there is no need to reinvent the wheel.

[27] R Wood and A Bandura (1989) Social cognitive theory of organizational management. *The Academy of Management Review* 14(3). pp. 361–384.

[28] JH Dyer, HB Gregersen and CM Christensen (2009) In the innovator's DNA. *Harvard Business Review,* December 1st.

Organizational Zoom . . .

Southwest Airlines: A Collaborative Community

"It's a community; we try to keep it a family of 35,000. It certainly requires different techniques to interact and communicate with 35,000 than it does with 500".

— Gary Kelly
President and CEO of Southwest

Since its inception in 1971 in Dallas, Texas, the world's first low-cost airline carrier, Southwest Airlines has captivated the attention of the business world. The company has maintained the momentum and agility of a young start-up company, has nourished a unique and powerful corporate culture, and has continued to grow and fluidly adapt to the changing environment year after year. In 2011, the Southwest Airlines celebrated its 40th consecutive year of profitability, while still focusing on the domestic market, serving 72 cities and operating over 3,400 flights per day, making it effectively the largest airline carrier in the United States. The company pulled through the worst of the crisis healthier than ever, with just one uncertain year in 2009, to demonstrate strong growth and profitability in 2010 without resorting to staff layoffs, which is an exceptional feat in an industry that is highly susceptible to the whims of the market.[29]

Southwest Airlines, as described by Gary Kelly, its President and CEO, is reputed for a highly collaborative culture where employees from all levels proactively bring their ideas for improvement to the forefront. Such a culture is central to the spread of knowledge between members so that the company can learn, innovate, and improve. Maintaining this valuable knowledge and expertise in the organization is the key. As Kevin Krone, Vice President Marketing, Sales & Distribution at Southwest Airlines underlined the importance of this valuable expertise: "*There tends to be a lot of tenure in the company ... so there is therefore deep expertise in the company which is very helpful... that helps immediately as you go through the [innovation] process because they know how things work. Those experts come in to solve the problem*".

The company's collaborative culture is sustained by a *community-ship* and collective endeavor. In fact, the company has a policy of making each individual privy to the information that matters in the company, thus making everyone feel like an important part of the company and thus creating transparency and accountability. This can be a challenge in a company like Southwest Airlines with over 30,000 employees.

[29] H Mathe, X Pavie and M O'Keeffe (2012) Innovation at Southwest Airlines: Reinventing the business model. *Innovation & Society* 23, June.

Yet, the culture is managed effectively through clear, transparent, and frequent communications, and is bolstered by the employees themselves. Colleen Barrett, President Emeritus of Southwest Airlines, explains how important this is: "*We encourage people to as long as they hold themselves accountable first, to then hold others accountable. As long as they do it in a respectful way*".[30] Jan Carlzon, the former CEO of SAS Airlines, echoes this sentiment stating that "*An individual without information can't take responsibility. An individual with information can't help but take responsibility*".[31]

Besides fostering the optimal culture for innovation, bolstered by the dispersion of knowledge, Mr. Kelly argues that the culture, which is characterized by an entrepreneurial and warrior spirit, is boosted by the intake of new people *"The culture thrives on adding new blood and new people"*.

Fostering a Culture that Supports Risk-Taking

Innovation can be a fuzzy process as it involves a set of intangible inputs such as creativity, imagination, ideas, inspiration, and motivation. It is difficult to quantify innovation into concrete outputs and results. It is a process that inherently involves a degree of risk. And yet, if taking risks with the unknown and the unquantifiable continually represent a barrier to innovation, the very risks that may lead to a "game-changer" will ensure that innovation simply cannot occur. Because the positive effects of innovation can only be perceived when there is a distinct surge in company growth. Executives who tend to use growth as the sole metric for innovation underestimate the value of it, while in fact; risk is inherent in the process and eventual success of innovation. We can find plenty of examples of companies that have experienced a seemingly dramatic failure in their product line, but have survived the failure.

Organizational Zoom ...

The Coca Cola Company

The Coca Cola Company famously reworked the legendary recipe for Coke with the introduction of New Coke in 1985 which was received with widespread condemnation leading to its recall after 79 days and a return to the original recipe. Despite taste tests with consumers, the "tampering" with the "sacred" Coke recipe was perceived by customers

[30] Interview with Colleen Barrett, President Emeritus of Southwest Airlines, September 1st, 2010.

[31] http://en.thinkexist.com/quotes/jan_carlzon/.

to be sacrilegious. Yet the failure somehow became a marketing success with sales of "Classic" Coke reaching new heights upon reintroduction.[32]

Today failure is considered a pillar of successful innovation in the traditionally risk-adverse Coca Cola Company. Indeed, the former Chairman and CEO, Edward Neville Isdell, tried to instill a more risk accepting culture, reassuring employees that "*You will see some failures. As we take more risks, this is something we must accept as part of the regeneration process*".[33]

Hence, apart from engaging people through recognition and reward, and demonstrating the value of innovation, the reaction of management to failed attempts toward innovation is also highly important. Innovation requires risk; it is a fuzzy concept with unknown outcomes. Without taking risk, innovation cannot be encouraged throughout the organization. Thus, how the management deals with the downside of risk-taking is critical for future endeavors in the domain of innovation. An employee who has made a genuine and calculated attempt to innovate, as far as possible, by introducing a new idea product or service which unfortunately fails on the market will lose all interest and confidence in further innovation if he is strongly reproached for this action. As Colleen Barrett, President Emeritus of Southwest Airlines, states, "*You have to be forgiving. We are very tolerant and forgiving when people make an honest mistake. You have to be very careful about how you approach that mistake, how you call it to the person's attention and how you discipline, if at all, and how you counsel, if at all*".[34]

Organizational Zoom . . .

Samsung: A Culture of "Crisis" to Stimulate Innovation

In theory, crisis should not be considered a permanent condition for innovation; it would not be a "crisis" unless it was a sporadic occurrence, rather than a continual state of affairs. Yet, in many cases, crises, downturns, catastrophes, and even a fall in profits can inspire the creativity and ingenuity required for successful innovation. But can the simulation of crises in the organization result in conditions that foster and sustain innovation? The CEO of Samsung Electronics, Jong-Yong

[32] Bloomberg BusinessWeek (2006) How failure breeds success. July 10th. http://www.businessweek.com/magazine/content/06_28/b3992001.htm.
[33] *ibid.*
[34] Interview with Colleen Barrett, August 31st, 2011.

Yun, has built a culture of *perpetual crisis* to stimulate innovation. Although the success of Samsung can be attributed to a plethora of strategic decisions, the company has undoubtedly reaped the rewards of an insecure company culture.

Jong-Yong Yun "*nurtures*" this culture meticulously which is described as a *"uniquely paranoid"* corporate culture, the one in which *"disaster lurks in every overlooked detail, and success only increases the danger of complacency and eventual failure"*. Yun relentlessly admonishes his employees with a series of fear-inducing scenarios creating the perpetual atmosphere that a crisis is just around the corner. He cites the threat of competitors, market implosion, and even potential supplanting of the company by China, in order to fuel the trepidation that drives his employees to try harder to cut costs and innovate.[35] In his monthly speeches to employees, Jong-Yong Yun habitually warns that Samsung is in a crisis as severe as the one that nearly destroyed the company in 1997, telling his staff that "*we now stand at the crossroads to becoming a world leader or a major failure*".

While Jong-Yong Yun's claims may in fact, be based on a genuine belief on his part, and not on a falsely simulated environment to ignite a fighting spirit amongst employees, nevertheless the level of innovation at Samsung is astounding. Samsung registers more patents each year (1,600 in 2004) than Intel and unveils "new to the world" products on a regular basis. While once Sony stood as its role model, today Samsung has doubled its market capitalization when compared to Sony and is the world's leading maker of memory chips, flat-panel LCD displays, and color televisions.[36]

Creating an organizational culture that stimulates innovation is even more challenging for a Korean organization bridled by traditional Confucian hierarchies. Although Korea has relaxed its ancient traditions as democracy has developed in the last fifteen years, respect for elders and a reluctance to speak out of turn are still the norm. Innovation often demands an insurgent element to speak out in the organization and challenge norms, which goes against the grain in a Korean organization. Yet, Samsung's designers have succeeded in breaking through the constraints of the traditional culture. In many parts of the company, Confucian order still prevails, but the design center has initiated a groundbreaking culture that sidesteps the norms. The design building is separate from company headquarters, albeit at walking distance, and it is here that a sense of freedom of expression prevails in the organization; there is no dress code, and everyone is encouraged to speak up and

[35] P Lewis (2005) The perpetual crisis machine. *Fortune International* (*Europe*) 152(4). p. 36.
[36] *ibid.*

> challenge their superiors. Amongst young people, hair dyed green or pink is not uncommon. Teams of three to five persons are the dominant work models for designers. Everybody work as equals despite different areas of specialization or length of tenure.[37]

Some companies play it safe and only innovate with products they are sure to sell, never testing the water for radical designs that have not been proven in the market or for which there is too great a degree of uncertainty. A tolerance for failure is necessary for companies that seek to push the boundaries between their safety zone and risk-taking. Risk aversion tends to be more common in SMEs due to financial constraints, fewer technical and managerial competences, and more limited access to information relative to a large organization.[38] Alberto Alessi, the President of Alessi, a company that has succeeded in delighting the customers with its unique and emotional designs, proposes a theory on this phenomenon for innovation. He refers to this as his theory of the "borderline" which divides the "possible" from the "not possible". "*Well-organized, mass production companies try to work as far as possible from the borderline. They cannot afford to take too many risks. But by all producing the same car, the same television set, and the same fridge year after year, those companies are making products more and more boring and anonymous*". On the other hand, a company such as Alessi that thrives on creating innovative design products tries to live as close as possible to the borderline, which enables the company to explore a completely unknown area of products.

As companies nowadays are becoming more conservative on their spending habits, today development teams at many companies are focused on reducing cost and creating line extensions or incremental improvements to existing products, rather than aiming to develop breakthrough products.[39] The authors Eric von Hippel, Stephan Thomke and Mary Sonnack mention that there might be two main reasons for this failure: Firstly, managers focus too much on the short-term, and secondly, they do not know how to create breakthrough ideas because of a lack of an "*effective system put in place, which can guide and support their efforts*".[40] At 3M, product

[37] BusinessWeek (2004) Samsung design. November 29th.
[38] D Brown (1997) *Managing Risk and Innovation: The Challenge for Smaller Businesses.* University of Warwick Risk Initiative Briefing.
[39] Video on The Most Innovative Companies. http://www.businessweek.com/magazine/content/10_17/b4175034779697.htm.
[40] E Von Hippel, S. Thomke and M Sonnack (1999) Creating breakthroughs at 3M. *Harvard Business Review*, September.

developers work within an innovation culture that supports risk-taking, in accordance with their aphorism "*It's better to seek forgiveness than ask for permission*", and were successful at creating innovative products. After a short decline in innovation in the mid-1990s, the company adopted the *"lead-user process"* approach, which was based on two major findings by researchers. Manufacturers are not necessarily the ones to invent new breakthrough ideas; it is often the user. Those users can be referred to as *"lead-users"*, since they are "*ahead of market trends and have needs that go far beyond those of the average user*".[41]

Instead of eschewing all possibilities of risky innovation, some companies have established programs that encourage innovators to "shoot for the stars" and take on projects that have a low likelihood of success. Royal Dutch Shell introduced the GameChanger program, which is designed to seek out and fund good ideas which are considered high-risk and have low prospect of coming into fruition, but which, if developed could have a profoundly positive impact on Shell's business. The stakes are higher and the rewards for successful development hold a much higher associated value than a low-risk innovation. GameChanger teams in each of Shell's business units examine, select, and support ideas throughout their development process. Ideas can come from within the company or from outside innovators, although Shell retains the intellectual-property rights to any designs or products that might be generated. The company devotes 10% of its innovation budget to the program, but the rewards are high: The program generates a full 30% of Shell's R&D projects. By recognizing the power of good but far-fetched ideas, Shell has reaped the benefits of a number of new technologies that only few thought were possible to achieve.[42]

Leadership that Fosters a Culture of Innovation

A study of innovation by the consultants Booz, Allen and Hamilton demonstrates the two chief characteristics which distinguish more successful innovative firms from their less successful rivals: Firstly, they manage to set a clear and effective direction for innovation that aligns with the firm strategy and strengths; and secondly, development is well managed with effective collaboration and a keen attentiveness toward the customer

[41] *ibid.*

[42] BT Ferrari and J Goethals (2010) Using rivalry to spur innovation. *McKinsey Quarterly*, May.

throughout the firm.[43] Southwest Airlines have a clearly defined list of innovation priorities that are communicated right through the organization to each and every employee so that ideas for innovation coming from employees will be on a common trajectory; instead of thousands of ideas for change and innovation spanning the organization, Southwest can focus on five key areas of development and communicating this with employees also conveys to them that they are a valued part of this development.[44]

But in innovative companies, what role does the leader actually play? A study by Dyer, Gregersen, and Christensen (2009) which observed the actions of 25 innovative entrepreneurs with over 3,000 executives and some 500 individuals who had started innovative companies or invented new products showed that executives at most companies do not feel personally responsible for coming up with strategic innovations. Instead, these executives felt responsible for facilitating the innovation process. However, in stark contrast, senior executives of the most innovative companies claimed that they do not delegate creative work; rather, they do it themselves.[45]

According to Teresa Amabile, one of the most efficacious things a manager can do to stimulate creativity and intrinsic motivation is "*the deceptively simple task of matching people with the right assignments*".[46] One of the most important roles that a leader plays in guiding innovation is person–job fit. Finding the right person for the right role is one of the prerequisites for ensuring that a person's skills are put to good use and that they are intrinsically motivated; being a key driver of creativity. Intrinsic motivation means that a person's motivation is driven by an interest in the job function itself, and research shows that this form of motivation plays an important role in stimulating creativity. Managers can match people with jobs that play to their expertise and their skills in creative thinking, and ignite intrinsic motivation.

In addition, clear direction and meaningful objectives give employees an understanding of the path that a company and its innovation strategy seek to take. Even it seems like leaders asking the impossible. The very existence of a clear objective gives employees the motivation they need

---

[43]D Midgley (2009) *The Innovation Manual.* Sussex: Wiley.
[44]Interview with Bob Jordan, Executive Vice President Strategy and Planning, Southwest Airlines. August 31st, 2010.
[45]JH Dyer, HB Gregersen and CM Christensen (2009). *Op. Cit.*
[46]T Amabile (1998) What kills creativity. *Harvard Business School Press*, September.

to be resourceful and in finding a solution to a complex problem. When Southwest Airlines found themselves in the situation where they had to sell one of the airplanes in their already small fleet to raise capital quickly, employees found themselves inventing the famous 15 minutes turnabout to solve the problem of fulfilling a customer promise, which was to be a reliable and frequent carrier. This particular process innovation revolutionized the industry model.

NATIONAL FRAMEWORKS FOR INNOVATION AND ENTREPRENEURSHIP

Over the course of the past fifty years, no single place in the world has stimulated innovation like Silicon Valley. It is not only the quantity of innovation produced in this area which covered nearly 3,885 kilometers, but also the fact that it has continued to produce innovation over time. And, as one study noted about Silicon Valley,[47] what makes it unique is its capability for developing new technologies and *exploiting* them commercially. So what is it about this particular area that makes it so conducive to innovation? How we take lessons on creating the right conditions for stimulating innovation from Silicon Valley?

A Supportive System or a Disincentive to Innovate

Private sector and governmental initiatives often play an integral role in generating opportunities for public–private partnerships, for providing stimulus and incentives to attract investments, and for creating a favorable ecosystem and conditions propitious to entrepreneurship and innovation. The question of whether it is in the national interest of a country to create a system that stimulates and supports innovation is a key question today and is a unique one for each country. Differences in language, political system, and culture can represent barriers for startups seeking to invest internationally. Even on the home turf, the rigidity of the system for the creation of new enterprise plays a key role in positively impacting the region's productive diversification, competitiveness, and economic development.

---

[47]CM Lee, WF Miller, MG Hancock and HS Rowen (eds.) (2000) *The Silicon Valley Edge. A Habitat for Innovation and Entrepreneurship*. Stanford, CA: Stanford University Press.

Investments in venture capital and private equity form a fundamental part of the national framework and support system for innovation. Such investment funds were fundamental in the establishment of the innovation hubs of Palo Alto (USA) and Tel Aviv (Israel). The resulting propensity for innovation in these countries has been greatly advanced through the incubation and stimulation of startups and leading industries in these countries. The USA and Israel invested comparatively better than other countries in entrepreneurial activities and industry innovation; this investment stood at 0.3% of GDP in the USA, for example, *versus* 0.03% of GDP in Latin America, 0.17% as the global average, and 0.19% as the European average.[48]

While national innovation levels can be difficult to measure due to disparities in measurement indicators, patents are one form of generally accepted proxy indicator, although they are arguably not the most appropriate indicator for measurement. One of the difficulties with patents is that certain innovations are less apt to be patented. While services accounts for some 80% of the economy in many Western countries, services innovations are often not patented at all, particularly in the form of incremental innovations. Yet technological innovations and product innovations rely heavily on the system of patents to protect an invention from being commercially exploited by another entity. Thus the international landscape of patent protection provides a foundational framework for the support and encouragement of innovation. Inadequate or complex patent regulation can be a major deterrent to would-be innovators. A patent provides protection of an invention for a period of time, generally twenty years,[49] and is therefore a vital incentive for the commercialization of a product or process.

Stimulating Innovation through Entrepreneurship

Entrepreneurship is widely recognized as playing a central role in driving the economic development of nations by creating new indigenous businesses, promoting job creation, stimulating innovation, and building economic and

[48] InnovaLatino: Fostering innovation in Latin America (2011) Published by Ariel and Fundación Telefónica, in collaboration with Editorial Planeta. http://www.innovalatino.org/documents/InnovaLatino%202011%20FINAL%20English.pdf.

[49] World Intellectual Property Organization. http://www.wipo.int/patentscope/en/patents_faq.html#patent_role.

competitive strength.[50] Indeed, since the beginnings of economic study, economists such as Smith and Ricardo have regarded entrepreneurship as a critical success factor for economic performance.[51] But does entrepreneurship play a role in stimulating innovation?

This has led to a growing interest in the development of education programs in higher-level educational institutions (HEIs) to encourage entrepreneurship. Education in entrepreneurship can serve to equip individuals with the requisite skills for launching a new venture or startup, and this acquisition of knowledge plays an important role in increasing self-efficacy[52] and achievement motivation,[53] factors that have been found to positively impact entrepreneurial success.[54] The conditions for setting up and running a SME or start-up company are less favourable in the EU than the US, according to a Euro-barometer study. The 2007 Euro-barometer survey found that EU citizens are less inclined to become entrepreneurs (45% *versus* 61%), and more risk-averse than their American counterparts. Starting with the education of entrepreneurs in HEIs, the European infrastructure for entrepreneurship leaves enough room for improvement. There are several well-established models of HEI entrepreneurship education in the US, such as MIT, Stanford and North Carolina. In comparison, no European model has reached the same level of maturity.

In October 2008, a study was concluded on Entrepreneurship Education under Higher Education (HE) in Europe. The study by NIRAS Consultants, FORA, and ECON Pöyry covered a total of 31 countries, including the 27 EU member states, Iceland, Liechtenstein, Norway, and Turkey. The study concludes that in most European countries the stated policy commitment to promote entrepreneurship has not resulted in making it a widespread subject in higher-education systems. While higher-education conditions for entrepreneurship have advanced to a mature stage in the United States, studies indicate that entrepreneurial education in European countries lags behind

[50]GR Plaschka and HP Welsch (1990) Emerging structures in entrepreneurship education: Curricula designs and strategies. *Entrepreneurship Theory and Practice*, 14(3). pp. 55–71.

[51]P Nijkamp (2003) Entrepreneurship in a modern network economy. *Regional Studies*. Volume 37, Issue 4. pp. 395–405.

[52]A Bandura (1986) *Social Foundations of Thought and Action: A Social Cognitive Theory*. Englewood Cliffs, NJ: Prentice-Hall.

[53]DC McClelland (1965) Achievement motivation can be developed. *Harvard Business Review, Nov–Dec* 7th.

[54]D Miron and DC McClelland (1979) The impact of achievement motivation training on small business. *California Management Review* 21.

that of the US and Canada. However, multiple initiatives have been taken all over the region leading to significant improvements by the mid 2010's.

Indeed, the study concludes that education in entrepreneurship is still in its infancy in Europe, and that institutions would benefit from a more formalized exchange of good practice in curricula and methods. Entrepreneurial education, a "learning by doing" subject, is more effectively taught when skill teaching is enriched with practical cases and real company examples, and the importance of establishing opportunities for collaboration and sharing of methods between European countries is underlined. As the US model of entrepreneurial education has predominantly inspired teaching methods in Europe, a strong motivation for the development of cross-border and cross-disciplinary collaboration between European countries exists, as there are arguably more valuable and more applicable lessons to be shared amongst EU countries than with the US.

Promoting entrepreneurial spirit in higher education is and will be one of the strongest priorities of the European Commission (EC). Financing entrepreneurial education initiatives in the HEIs is enhanced by the use of EU structural funds; the initiation of EU programs dedicated to entrepreneurship; the promotion of entrepreneurship and technological transfer in research projects funded by the EC. The HE sector is seen as playing a strategic and responsible role in local economic and social development. In the context of greater uncertainty and complexity for business, the entrepreneurial paradigm must evolve.

Innovation Success and Country of Origin

When does a corporate brand reflect a nation, and when does a nation reflect a corporate brand? Coca Cola has been built globally upon its synonymy with America and the American dream; IKEA reflects the sense of quality associated with Sweden, fashion is intimately linked with Italy, and good food with France. Such branding can even have a bidirectional impact whereby companies can lean on national brands to reflect their defining qualities and in turn the national brand may be boosted.

While a strong corporate brand image can boost or hamper a nation's image in the eyes of the global market, national image can bring with it a propitious foundation or a destructive association for corporate development. When Starbucks was first launched in France, it was perceived in a negative vein much the same as McDonalds, i.e., as an invasive American corporation that would upend the indigenous service industry in France. Despite such inglorious beginnings, Starbucks has managed to redefine customer perception and establish its chain of coffee houses as a quality experience. While IKEA was founded on Swedish credentials of quality and clean design, today it could be argued that IKEA actually does more for the country's image than the tourism department. Ollie Wästberg, Director of the Swedish Institute wrote that "*IKEA is doing more for the image of Sweden than all governmental efforts combined... IKEA's 285 stores in 37 countries feature the blue-and-yellow national colors, serve Swedish meatballs, and sell blond-wood Swedish designs and books about Sweden. To visit IKEA is to visit Sweden*".[55]

For the corporate world, the equity and value created by the national reputation are equally important for the development framework. Germany is linked with engineering prowess and excellence; qualities that Opel capitalizes in its marketing campaigns by using the German tagline "Wir Leben Autos" (literally "We live cars") on a global scale. The German only tagline reminds customers of the German roots and place of manufacture; although it is General Motors, the second best-selling brand. Alain Visser, Vice President of Opel Marketing, Sales and After-Sales confirmed this strategy: "*It's a gutsy move to stick with German instead of going with English. I think just the choice of language itself already makes a clear statement*: *We are an international company with German roots*".[56]

The importance of a good foundational national brand is highlighted in South Korea's struggle to differentiate itself amongst fellow Asian countries. In 2006, a survey by Anderson Analytics of 1,000 US college students across 375 schools revealed that only 39% believed South Korea produced quality products, while 81% thought Japan did. Unfortunately the successful brands such as Samsung and LG were not identified as South Korean. In fact, less than 10% of the students in the Anderson Analytics survey identified Samsung or LG as South Korean. The students believed that such brands were Japanese. Thus while South Korea has worked to develop a state of the art manufacturing system, Japan has somehow taken the credit due to unclear identification with the product.

---

[55]C Graves (2010) South Korea: Finding its place on the world stage: Designing a distinctive national brand. *McKinsey Quarterly*, April. https://www.mckinseyquarterly.com/Economic_Studies/Country_Reports/South_Korea_Finding_its_place_on_the_world_stage_2577?pagenum=3.

[56]Autoevolution. Frankfurt auto show: The new Opel is born. September 15th, 2009. http://www.autoevolution.com/news/frankfurt-auto-show-the-new-opel-is-born-10932.html.

A survey by the South Korean Presidential Council on Nation Branding found that consumers in Japan, Germany, and the United States marked down the value of identical products by 30% or more when told they came from South Korea. This pricing discrepancy is evidence of the power of a nation's brand. Negative perceptions of the country of origin contribute to the perception of quality of a product.[57]

Indeed, people are the only asset that innovates, and innovation is the only path to sustained breakthrough performance. To unlock the innovative potential of its talent, we suggest that companies should leverage the five drivers of individual innovators' motivation, and create an environment where employees can acquire knowledge and skills, connect with others, well comprehend the strategy and vision of the organization, defend their identity with the organization, and have clear direction and management. On the organizational level, companies should foster a leadership and a culture of innovation that support collaboration and risk-taking. Finally, it is also important to develop a supportive system on the national level to propagate innovation and entrepreneurship. The later topics will be developed in a later chapter of this volume.[58]

[57]C Graves (2010). *Op. Cit.*

[58]Ziying Huang, Pascal Minella, and Marwyn O'Keeffee deserve much credit for their assistance in the writing of this chapter.

Chapter 8

GREATER SPACE MEANS BETTER LIFE

Administrative structures are certainly of great importance in order to determine employees' behavior and performance. On the other hand, physical structures also significantly influence the way staff and customers view the organization and interact with it. Competitive economy firms and institutions are competing due to innovations in products/services, delivery processes, and management styles, which is the present scenario observed in creative environment in the contemporary service/experience. Innovative approaches frequently materialize in the design of service facilities through which experience is provided. Despite the massive deployment of the digital transformation, physical structures in which physical operations are conducted still play a key role, and architecture still generate emotions when delivered through the screen of a computer. Organizations are in a position to significantly improve convenience, productivity, engagement, and attractiveness by carefully designing space and defining the appropriate layout. Facility layout can be addressed with attention to traffic flow, space planning, and the need to avoid unnecessary travel. This pattern also has to include identification of the meanings, characterization of size and qualification of the process by which any facility delivers its messages.

ARCHITECTURE AND THE BUILDING OF CORPORATE SYMBOLS

Organizational Zoom...

HSBC versus Bank of China:

Historically, the dramatic Hong Kong & Shanghai Bank tower on Victoria Island in Hong Kong which was designed in 1979 by Norman Foster certainly symbolized the economic success of this small colony based on free market dynamics and true managerial attitudes. Foster's achievement reigned supreme on the skyline above the whole City until the Bank of China, an institution from the People's Republic, built an even taller building immediately behind it. Since then, both buildings have been dwarfed by the erection of numerous taller additional skyscrapers in the area.

This way, by competing for the tallest building in Hong Kong, during the 80s, two different power groups, conveying two opposite political ideologies, worked very hard to give out strong messages to the local population as well as to foreign observers. Beyond the competition for the most prominent position, the towers themselves were designed in order to convey different messages: The completely open space of the HSBC facility *versus* an enclosed one for the Bank of China; an external and visible bright iron structure for the first compared to a traditional structure hidden behind brown facing for the second; an exuberant but balanced shape on the one hand and solemnity and dissymmetry on the other. These pieces of corporate architecture gave visitors and users, messages about what the two successful giant financial companies believed in, and what and how they achieved their work. In order to communicate their view, both buildings used size, but differently due to the chronological order of their construction as well as the field space available at the time, and also formality, but at two very distinct levels.

Successful Facility Design for Organizations

Operations can be directly affected by the facility design.[1] A restaurant with inadequate ventilation for non-smoking diners may discourage many customers. A physical fitness center that has easy wheelchair access may be able to enlarge its services to include a new clientele. Design and layout represent the supporting facility component of the service package. Together they influence how a service facility is used and, sometimes, if

[1]H Mathe (1997) *Le Service Global*. Paris: Maxima.

it is used at all. Consider the dramatically successful Canadian Shouldice Hospital. A good portion of its success in repairing internal hernias resulted from thoughtful design and layout. For example, the operating rooms were grouped together so that surgeons may easily consult with each other during procedures. Because early walking promotes faster healing, the hospital was designed to provide ample pleasant places to walk and even to climb a few steps. Meals were served only in community dining rooms rather than in patient rooms. This requires more walking and as an added benefit, allowed patients to get together and compare notes.

Organizational Zoom...

Wal-Mart versus Kmart:

In several surveys conducted during the early 90s, customers' perceptions of the two leading American retailers show significant differences. Shoppers viewed Wal-Mart, and actually still do, with higher esteem than Kmart, while neither merchandise nor prices differentiate the two. Close analysis unveiled differences in store decor and layout. At Wal-Mart, the main aisles were wider than those at K mart. Fluorescent lighting was recessed into the ceiling, creating a softer impression than the glare from the exposed fixtures at Kmart. The apparel departments were carpeted in a warm, autumnal orange, while Kmart's were tiled in off-white.

Together, such features signal consumers that Wal-Mart was more upscale and that it carried merchandise of a little better quality than Kmart's. Wal-Mart's attention to facility design details has helped shape shopper's attitudes by constructing the delicate balance needed to convince customers that its prices are low without making people feel cheap.[2] Wal-Mart has successfully used facility design, to differentiate itself from its competitors. Using facility design as part of a differentiation strategy is now very common.

In Europe, investigations unveiled significant differences regarding the amount of energy that companies are allocated in this architectural field.[3] The most advanced ones view space design and management as strategic resources, while most of the other still seems to feel uncomfortable at taking

[2]Schwadel, F. (1989). Little touches spur Wal-Mart's rise. *The Wall Street Journal*, September 22.
[3]M Dubosson-Torbay (2000). *Gestion des Espaces de Transaction de Service et AvantageCompétitif*, PhD research dissertation under the supervision of H Mathe, University of Lausanne.

advantage of the open possibilities. However, sophisticated methods and tools have been developed and could be easily mobilized. For instance, the concept of process flow analysis used by industrial engineers may be adapted in order to appraise layout of working facilities. Also, analysis of settlement layout, using the generative syntax method allows for establishing a description of spatial order.

Shaping the Contour of Corporate Space

On top of influencing productivity and convenience, buildings convey messages. In providing emotion, and because of people's sensibility to their environment, space certainly means, or at least, evokes multiple feelings. When space is the result of an effort by an architect, it probably has to mean something positive. This is the opinion of Sir Herbert Read when he affirms, "when there is no symbol and, therefore, no discourse, there is no art". Architecture certainly carries messages about itself; the creative process adopted as well as the cultural influence recorded. It may also carry external messages as a main purpose, and in a more or less aesthetically valuable way. The question we must immediately raise after accepting this proposition is, "what does space mean to people?" If we consider just the basic categories of observers, customers and employees using a building, architects, critics, and owner or developer, there is a serious probability that the meaning of a given space would appear differently to each particular group. Another question might be posed: How long does such a "space" exist?

Bill Hillier, from University College London, asserts that a theory of architecture does not say what should be done, which to some extent is being a moral and political issue; but rather how to do it.[4] Following his assertion, the validity of the messages that company management wants to communicate do not have to be evaluated here. The fact is that influencing customers, investors, and employees' behavior is basically what management is all about. Starting with the hypothesis that architecture encourages and/or discourages certain behavior, especially in service activities given the intangibility of most of service products; the relevant question is rather "in which direction do corporate leaders try to take advantage of this resource when they do so".

---

[4]B Hillier (1996) *Space is the Machine.* Cambridge: Cambridge University Press.

Among others, it seems that perhaps the most significant trends lie in the following:

- *The reinforcement of ideology.* Therefore, design expresses the necessity of faith. For instance, the Battersea power plant, built in London in 1929 and recently converted into luxury mall and office quarters, strongly communicates the faith in the new scientific and technological dynamic which inspired all the work carried out under the notion of progress during the major part of the 20th century.
- *The reinforcement of individual or an organization's position.* Design displays success and power earned. Kevin Roche, designer of the controversial General Food's headquarters in New York's Westchester County, was chosen to conceive Challenger, the headquarters of the Bouygues Group in France near Paris. The brochure later produced to publicize this modern palace, started as follows: "Francis Bouygues, founder and chief executive of the Bouygues Group, leader of the international construction industry, sought to unite all branches of the Group's activity at Challenger, a head office on a scale unprecedented in Europe".
- *The reinforcement of social rules.* Thus design invites people to follow conventions and given practices. The carefully designed layout at any IKEA commercial facility allows customers to gradually discover the Swedish retailers' products through a series of exhibition stages before naturally finding themselves in the warehouse where the products can be collected, purchased, and easily transported to their cars. Most Club Med villages have also been designed in order to convey a sense of community and to foster a certain kind of behavior. The layout invites people to converge to the center of the facility where food is served and entertainments are organized.

A very strong body of ideologies dramatically affected Society all over the world during at least two thirds of the 20th century. Socialism, democratization of a nations' power as well as egalitarian and non-discriminative practices are all addressed. Characterized by uniformity and repetition of very pure and basic elements and using raw materials for facings, the so called rigid "International Style" communication through design did not necessarily match the requirements of corporate leaders who were traditionally comfortable with the rich and pontifical bourgeois turn of the 19th century architecture. Meyer, Mies Van De Rohe, Melnikov, and later Le Corbusier or Niemeyer amongst others imposed the messages of International Style the world over, facing much less resistance in

Continental Europe than in the UK and the US, where they had to take more account of the client's requirements and wishes. In New York, both the Seagram Tower designed by Mies Van der Rohe in 1954 and the Pan Am Building by Gropius and TAC in 1958, illustrate the start of a transition period offering new opportunities for company management to communicate messages regarding corporate "ideology" or, more simply, system values. Even the basic skyscraper shape symbolizes a very strong statement to employees — reach the top! Seiler, who was teaching at Harvard during the 80s, proposed the following statement: "Management can set out to create environments that augment strategies both structurally and aesthetically. Indeed, to fail to do so is to waste a valuable organizational resource".[5] "Should size and formality be used in such an effort, can they be tapped on to foster the expression of imagination?"

The Size of the Working Building

Organizational Zoom...

Chicago Tribune:

In 1922, the board of directors of the Chicago Tribune called upon architects to submit plans for the no less than the "greatest building in the world". Five years later, Van Allen had to invest in creative solutions to raise the height of the Chrysler Building in New York as much as he could, because the latest city projects were for construction of yet higher skyscrapers, such as the Empire State Building and the Rockefeller Center, which seriously affected the mood of his own client.

Multiple cases show that size is used to communicate messages when it comes to buildings. It is particularly evident when both the CEO and the architects collaborate in order to create publicity; thanks to the building. Regarding space, Abercrombie sub-consciously writes, "We judge the object's size, using our own size as the measure. Is it smaller than we are, or bigger? If bigger, how much bigger?"[6] Beyond this primary reflection, an important point has to be made; "greater means more" should not be understood as, "greater means better". Twenty million cars obstructing the streets of London are certainly not better than two million. "Our house is like a convention center compared to the other houses here", said David

---

[5] J Seiler (1984) Architecture at work. *Harvard Business Review*, Sept–Oct.

[6] S Abercrombie (1984) *Architecture as Art*. New York: Harper & Row.

Siegel, the founder and chief executive officer of Westgate Resorts, the world biggest private time-sharing company, describing it's never-finished 9,000 square-meter dream house in Florida that he and his wife named "Versailles". Built on a custom-made hill, the huge mass of questionable concrete and marble achievement occupies one full acre and was supposed to be the largest home in the United States. Faced with business downturns in 2008, Siegel suddenly found himself $1.2 billion in debt, with no savings, and had to stop the oversized project at an advanced stage of development.[7] Following Abercrombie, we should agree upon "part of becoming a civilized adult, after all, is learning the lesson that quality is independent of quantity, and even that quantity may conflict with quality".

One could question how prominent landmarks affect the balance and harmony of the surrounding environment. Sometimes the repetition in a sequence of strategic locations of a relatively small, but very characteristic space item may also provide a feeling close to the emotions stirred by huge pieces of architecture. On the other hand, great space could sometimes result from the accumulation of additions to an initially modest building. For instance, this is the case in the historical and dramatic manufacturing site of the Renault Automotive Group which used to occupy the whole island of Seguin in Boulogne-sur-Seine near Paris for dozen of years.

A great space could also be required simply to satisfy functional needs even if the company does not want to impose its presence in a given environment through a drastic modification of the initial harmony. In such a case, messages conveyed are probably different from those displayed by the General Food or Challenger concepts. However clients may call upon the same architect to perform a completely different design. Corning Glass asked Kevin Roche to design a building "consistent with the foot print of Main Street", Corning, New York. Its new centrally located headquarters paralleled the nearby three-story structures as well as echoed their materials, so that it fits into the block like an "old shoe". Hence the way a space is handled and the level of formality it desired, partly determine the message communicated.

Process to Influence

"Within any building size, what a myriad of shapes are possible!" explains Abercrombie. "The design of those shapes is a more subtle and a much more

[7]S Berfield (2012) The Siegels of Versailles. *Bloomberg Business Week*, March 15th.

complicated way of affecting our response to architecture". Formality may be requested to include space in order to communicate messages linked with ideology, position or social rules. In fact, traditionally formality was used in architecture, education, religious or military rituals, etc, to express the views and expectations of the authorities. Formality also reassures people of the reliability of the relationship established with these authorities. The complex processes of communication and influence may be divided into different types of distinct, but potentially convergent, basic processes. Among others, three such processes are easily identifiable: The selective, the learning, and the mobilizing process. The selective process results in the differentiation of people and the manifestation of social scale. The learning process allows people to increase their own knowledge and abilities as well as the involved social organization. The mobilizing process tends to focus the energy of each individual on the improvement of group performance.

An incursion in the manufacturing environment brings value to the discussion. In the case of the New England factory for Teradyne Connection Systems designed by Huygens & Di Mella a few years ago, the company's management wanted the "assembly crisis problem", an unfortunate feature of the old plant, to be solved by the new facilities.[8] This obviously impacted the layout of departments and influenced their character. Therefore, the architect proposed that the assembly area be considered as a focal point of the building, surrounded by the specialist departments. "The new building made it difficult for the specialists to avoid involvement". In choosing such an approach both management and design exploited the opportunity of influencing behavior; thanks to space configuration. Distinction between workers on the line and specialists remain, but specialists now appear much more as "servers" of the assembly area. Being permanently connected to the line, the specialists teach the workers how to do things better, and they also learn from each other when they intersect at the focal point. Design is the key factor of such an improvement, and in this case it improves the consistency of efforts and mobilization of energy around production and quality objectives. Rather than using formality as a means, the new shape emphasizes direct and casual relationships between people, due to the organizing circulation systems.

[8]Seiler J. (1984). *Op. Cit.*

Serious works have been achieved over the past twenty years, especially in the service sector, to design corporate buildings that emphasize well-being and a casual relationship rather than rigidity and solemnity. This is even being applied to really large spaces. In many fields of business activity, success factors change as do communication requirements. What might have been good yesterday to improve attractiveness and productivity may no longer apply today; so creative solutions have to be elaborated. As huge pieces of a common and very particular space, the Disneyland and Disneyworld parks are created to communicate special feelings but certainly not formality. The Californian retailing chain Best is now well known for its series of extraordinary shapes, designed by SITE and Wines, which since 1977 have emphasized humor by shaped, completely unbalanced structures, and unexpected elements. In ordering the design of the BCV support center in Prilly, near Lausanne, or the SNCF freight headquarters in the rue Mouchotte in Paris, management shows that it pays attention to restoration of a certain area of comfort and of a cozy atmosphere in administrative buildings by using green space as a focal point. The emergence of campus type corporate buildings seems to come from this same consideration coupled with a respect for the environment. Describing the Becton Dickinson headquarters in New Jersey, designed by Kallman McKinnell & Wood, *Architectural Record* reported that, "the client wanted a corporate office building that looked as though it could have been built any time in the last one thousand years".

Buildings communicate their messages regardless of the intentions of their owners or users. Building greater space means enlarging the size of the conveyed messages which could eventually lead to increasing the influence on social behavior. Architectural historian, Christian Norberg-Schulz asserts that human life is largely conditioned by the quality of its "existential space", which itself affects the image of the environmental structure. Companies belong to such a structure where they are major social entities and fundamental elements of the environment. As such they must fully assume the responsibility related to their spatial integration in the milieu. The responsibility of the architect might be to teach corporate management how to do it.[9]

---

[9]T Walton (1988) *Architecture and the Corporation. The Creative Intersection.* New York: MacMillan.

LEVERAGING ARCHITECTURE TO TRANSFORM ORGANIZATIONS

Virtually almost all business activities, especially in service sectors, rely on buildings to provide its value to the market. Actually in hotels and restaurants, for instance, the core of the offering could be described, in simplistic way, as the provision of a roof for the night. Eating also necessarily takes place in some sort of physical surroundings. More and more, restaurants and fast food outlets are being located in commercial mall, train stations, airports terminals, and like all large physical facilities in which only service businesses stand. Banking corporations have always put great emphasis in conveying messages through the statures of their headquarters and branches; and most of the back office jobs are located in large office buildings. Government institutions and religious organizations have showed even more interest for mobilizing architecture as the way to communicate. Middle age Cathedrals served the very purpose of telling the official story of Christianity and of gathering the all population of a given area at regular dates for reinforcing social rules.

Innovative Space Design and Competition in Hospitality

In 1993, the first Hyatt Regency Hotel in France opened in Roissy-en-France, immediately close by Paris Charles de Gaulle Airport. Designed in an almost hostile physical environment by the Chicago-based German architect Helmut Jahn and the Parisian team of Jean-Marie Charpentier, the original, almost provocative facility is primarily organized in two austere wings. The two wings are linked to each other by a glass roof of unusual proportion: 40 meters in width and the equivalent of a seven floor construction in height. This astonishing piece of glass architecture covers a large lobby-restaurant-bar decorated with wood and shiny color panels. In front of the glass facade, but still belonging to the courtyard between the two wings, a park unveils a series of water ponds carefully positioned between several lines of trees, offering the hotel its own manicured landscape and sight. None of these features can possibly be anticipated from the nearby motorway.[10] Since its launching, the Hyatt Regency at Roissy has been recognized as a key landmark in the area, and successfully houses multiple activities, conventions, and events. This case seemed unusual in hospitality for early 90s, but one had to remember that the Pritzker family

---

[10]M Champenois (1993) Hôtels long-courriers. *Le Monde*, February 13th.

from Chicago, founder of Hyatt has endorsed a positive approach toward innovative architecture as a key strategic resource from the early age of the hospitality group.

During the late 60s in North America, while most players in the industry almost everywhere concentrated on the reproduction of cost effective, low ceiling, standardized practical structures, the Hyatt group invited John Portman to design its first atrium-centered hotel. Portman, who was credited for having beautifully renovated Atlanta downtown, was fascinated by Jules Vernes or Piranese, and was a great admirer of Mies Van des Rohe and Saarinen. He has been working on a new architectural concept for hospitality building for almost a decade. His original piece was allowing all the guest rooms to surround a dramatic atrium tower lobby on all 21 floors, with balconies and plants everywhere in the lobby. For the first time the elevators were engineered to be built with three faces of glass. No corridor was included in the drawings. For the heavy players in the hospitality industry at that time, such a design was considered far too expensive and generating too much unusable space. A small company launched in 1957 specializing in airport lodging services decided to embrace the innovative concept in 1967 and then, Portman and Hyatt working together literally created an entirely new style in hospitality architecture. The challenge to hotel architects was no longer to eliminate extra space, rather to create grand, wide-open public spaces. By 1969, there were 13 Hyatt hotels in the United States and the first Hyatt International unit was launched in Hong Kong that year. The hotel in Chicago that opened in 1971 received excellent appreciations, followed by San Francisco in 1973, then Atlanta, Detroit, and so on. Later on, Don and Jay Pritzker founded a special recognition award for innovative architecture: The Pritzker prize. Many observers in the building trade would consider this award in architecture equivalent to the Nobel Prize in science or literature.

Organizational Zoom...

Hyatt hotels:

Listed by Fortune magazine as the 95th-best US company to work for in 2014,[11] the Hyatt Corporation worldwide operates several chains and

[11]http://money.cnn.com/magazines/fortune/best-companies/2014/snapshots/95.html?iid=BC14_fl_list.

its portfolio consists of more than 550 properties.[12] Previous to the acquisition of Summerfield Suites in 2005 and the launching of the Andaz brand in 2007, and more recently of all-inclusive resort brands Ziva and Hyatt Zilara, the company operated luxury hotels and resorts around the world, including a majority of units in the US, Canada and the Caribbean managed under the Hyatt Hotels Corporation umbrella, and a series of hotels and resorts in 40 countries through Hyatt International Corporation. In 2011, the company entered Asia with Andaz Xintiandi Shanghai, with the newest of this new series opened in Tokyo in 2014.

The hospitality group enjoys a strong reputation not only for the physical distinctiveness of its buildings, incorporating local art and design, but also for the amenities and services provided.

On another dimension, Hyatt, which employed 80,000 staff members worldwide, is credited for efforts in diversity; the US arm leadership on the subject has led Fortune magazine to name the company as one of "America's 50 Best Companies for Asians, Blacks, and Hispanics" in 2004,[13] and in 2012, the Human Rights Campaign awarded the company 100% in the HRC Equality Index for eight consecutive years.[14]

As expressed by George Orwell, hotels were traditionally glorious illusions,[15] and hospitality has been generally perceived as a conservative sector of the economy, in terms of management style, attitude toward customer or guest, as well as facilities architecture and interior designs. Multinational groups, such as Accor or Marriott have reproduced hundreds of similar Novotels or Courtyards design. Until recently, staying at the Novotel in Sydney was producing an almost similar experience as the one at any Novotel in Cergy Pontoise, 25 kilometers North West of Paris except for the outside view from guest rooms; and the Courtyard at Roissy Charles de Gaulle Airport roughly relies on the same layout and on the same furniture set as any of its sister-units in America. Similar classical, if not Victorian, sofas, beds, chairs, and tables could be found in most places and menus will usually include such items like continental breakfast, pizzas, and pastas on room services, cheap-brand scotch and Coca Cola or Pepsi in the mini-bar always accessible in the bedroom with a special key for whatever unmentioned reason. Given the combination of good analyses of the average customer expectations and cost effective management requiring

---

[12]Company's Web site in 2015: www.hyatt.com.
[13]Special Report on Diversity in the Corporation. (2000). *Fortune Magazine*, July 10th.
[14]Our Brands. Hyattpressroom.com. Retrieved October 10th, 2012.
[15]The Economist (2013) A short history of hotels, be my guest. December 21st.

scale economy, all these practices make sense, even if participating toward this tendency of world "uniformization" that many fear, while considering the growing globalization of the economy.

Fortunately, the rise of "design hotels" opened a new and far more creative era that can be traced to 1980s New York, when Ian Scharger and Steve Rubell opened the cool and different Morgans boutique hotel. The idea was to pack bars and restaurant spaces "with locals rather than being cross-subsidized ghettos for lonely executives".[16] In the 1990s, the Starwood group followed the path and created the W chain, with the goal to redefine "the luxury and design-led lifestyle hotel experience", the most dramatic success in hospitality in four decades. Developed by British group IHG, the first Hotel Indigo opened in Atlanta in 2004, as "a brand hotel with design cache and service personality of a boutique". As of June 2014, the portfolio has grown to 58 properties around the world, including units in Hong Kong and Shanghai.[17] The rise of the design hotel may have to do with the general growth of travel. A 2014 Oxford Economics study, commissioned by solutions provider Amadeus, forecasts a 5.4% annual growth over the following ten years, outpacing global GDP by 2%.[18] No place is left unattended. In 2010, St Regis Lhasa resort, a partnership between Starwood and Lhasa Yungo International, opened in the capital of Tibet, featuring a five-star spa,[19] while Shangri-La opened its own luxury hotel in 2012 with Tibetan welcome rituals and butter tea. Inter Continental Hotels Group is supposed to provide its own addition in 2015. Finally, luxury camping spots, sometimes with safari-style tents like at Paws Up Montana's Resort, are also blossoming from North America[20] to Kenya and to Ladakh. Hundreds of examples could be added, often connected to individual initiatives by privately owned hotels. For instance, in the late 90s, Olivier de Vleeschouwer, an author and books editor, has established a list of recently built or restructured hotels for which he found the designs amazing. The collection include, among other creative expressions: The

[16]The Economist (2013) A short history of hotels. *Op. Cit.*

[17]C Huang (2014) Design hotels, hospitality chameleons. *Business Traveller Asia,* November. pp. 491–504.

[18]*ibid.*

[19]N Brandt N (2010) Luxury hotels reach new heights in China. *Bloomberg Businessweek,* December 6th.

[20]Falling Waters Adventure resort in North Carolina, Rock Creek ranch at Philipsburg in Montana, and Camp Orenda in the southern Adirondacks, among others in the US; Camp Thiksey in the Indus Valley, the Arctic Safari Camp in northern Canada, or Kohima Camp in Nagaland.

Sorat Art'otel in Berlin and the Pflaums Posthotel in Pegnitz, both in Germany, Il Palazzo in Fukuoka and the Poluinya in Hokkaido, in Japan, the Saint-James in Bouliac and the La Pérouse, in France, the Mondrian in Los Angeles, US, and the Pousada de Nossa Senhora da Asuncao, Arraiolos, Portugal.[21]

Imagination also seems to express itself in even most spectacular and lively ways in trendy restaurants and bars. At the 2003 Equip' Hotel edition, the trade fair in Switzerland revealed new trends in hospitality, and the visitors discovered a bar of large dimensions designed by architect Marion Bottier. When asked to describe her intentions, she said: "Genuinely modern, this bar aims to reflect the new trends in consumer spending behavior. Voluntarily gigantic, built on a monumental scale, we designed it in elevation (nine meters high). That brings it the optimal visibility. We installed it in two floors with a bar space on the ground floor and DJ station on the second (...). The décor enhances the key ideas behind the concept and translates an enchanter universe. Sharp colors are following crazy curbs to confound the visitor who will be brought to walk across sizeable objects. All of this is experienced in a multicolor light contributing toward enhancing the beauty of large fruit baskets and coffee bean bags. It is clear that customers are not going out today only to open up the bottle they could find in their own fridge; we must then astonish them, blow their mind and invite them to discover new senses".[22] In 2009, Tao Las Vegas generated more than $59 million from its restaurant and lounge. Looking like an oversized temple, the 6,200 square-meter restaurant displays a 20-foot-tall Buddha as well as a pool filled with Japanese carps, and hires people whose sole job is to put on a bikini and sit in a bathtub filled with roses.[23] It was by far the highest-grossing restaurant in the US. In 2014, the annual Restaurants and Bar Design Award toasted the world's most stunning places. From more than 860 entries, the judges selected two winners: Les Haras, a brasserie in Strasbourg, France, designed by Jouin Manku, and FEI, an upscale watering hole located in a W hotel in Guanghzou, China, designed by A.N.D.[24] A beautiful environment cannot, however, compensate for poor service or lackluster food, but it certainly

[21] O deVleeschouwer (1998) *Hôtelsétonnants.* Paris: Telleri.

[22] Ehlite (2003) With studios decoration, a special report on 2003 Equip' Hotel (EcoleHotelière de Lausanne), no. 3, July.

[23] J Stein (2011) Welcome to the Soy luck Club. *Bloomberg Business Week*, January 3rd.

[24] B Lanks (2014) Design on the menu; the world's most stunning bars and restaurants. *Bloomberg BusinessWeek*, October 2nd.

plays its role at attracting sophisticated consumers. Speaking of patrons, the bankers and lawyers of Lower Manhattan have since started to enjoy eating at Hudson Eats, a well-designed high-end food hall of 14 eateries that entirely rewrites the scenario for food courts.[25] Similar markets have flourished in Tokyo, Paris, and San Francisco.

In Switzerland again, a country usually noted for its traditional culture and quiet conservatism, the hotel Palafitte might have earned its rights for a durable life due to its originality. Initially built as an ephemeral artifact in Neuchâtel for accommodating the VIP visitors of the National Swiss fair "Expo.02", the unique luxury hotel Palafitte comprises of 40 individual lodges built in wood, including 24 of them standing on posts above the water directly on the lake and facing the Alps Mountains. In 2002, the Sandoz foundation poured Swiss Francs 20 million in the creation of this creative five star hotel that was expected to have disappeared by March 2004.[26] Due to its success even after the ending of the Expo.02 fair, and to the quality of the aesthetic impression produced on the neighboring area, the local authority has allowed the attractive establishment to survive the critical date, at least for some time. And it actually served as a laboratory for innovative, water based, hospitality facilities.

When advertising for the services they provide, companies permanently use architecture and sometimes architectural innovation. Among the noticeable campaigns, one might recall the impressive 2003 InterContinental effort. InterContinental hotels and resorts chain has not been very visible in magazine advertisements for the last decennia. After a somehow trouble period and many months invested into the redesign of its services and facilities, the group launched a strong and stylish advertising based on a concise message and a bold picture of the InterContinental London. Usually introduced on a full double page format and framed in black ink, the relatively dark picture featured a very high ceiling lounge, structured by pillars, open through a glass bay on a dramatic early morning view of park and city as seen from a higher floor, i.e., a higher level. In the middle of the right page and in the back of the room but close to the large window stands by himself a discrete executive holding a cup of coffee. On the wall behind him, almost in the darkness, one might have been able to see five drawings of classical architecture, all of them representing gates. Printed in

[25]M Reyes (2014) Don't call it food court. *Bloomberg BusinessWeek*, September 4th.
[26]S Jubin (2003) Neuchâtel se bat pour que le cinq-étoilesPalafitteéchappe à l'éphémère d'Expo.02. *Le Temps*, January 17th.

large white letters, the main message said: "It's easier to think outside the box when you're not staying in one", and a second one, almost invisible at first sight, added the following: "You can rarely mix business with pleasure. But that doesn't mean that business travel can't be pleasurable. We know what it takes".[27] Only a Web site address and a phone number in the bottom left page completed this creative ad. Actually this campaign has proved so successful that it was reused by WPG, the World Press Group whose member publications include BusinessWeek, The Economist, Fortune, Herald Tribune, Newsweek, and Time, and it advertised on its own capability to cost efficiently cover up-scale business people across Europe. A copy of the initial ad covers about half of the page delivered by WPG, under which a new message says: "InterContinental had designs on CEOs. WPG delivered them. Here's the proof". The text beneath the title summarized the communication strategy of the hospitality group and quotes some unnamed top executives: "The response from the international press advertising has been phenomenal. We deliberately moved away from stereotypical hotel advertising and our fresh-thinking, updated strategy has paid off handsomely".

Examples of successful design of trendy restaurants in large cities all over the world, starting with London, New York and places like Shanghai or Singapore, as well as a new generation of independent and creative hotels also show how much value a sound and innovative architectural strategy can provide to open-minded entrepreneurs. Is the lodging industry and restaurant business sufficiently mobilizing the resources for innovation provided by smart architecture? The like 1970s Hyatt determination to break with the design practice of the sector is still rare. Hotel and resorts corporations, restaurant chains, and entertainment parks, cruising companies and casinos, are obviously investing large sums of money into the building and the maintaining of their facilities. Financing any new piece of real estate is carefully thought about; strategic planners often manage to put together pools of investors who are willing to share risks and rewards. However, not enough of these firms have top executive focusing on leveraging the innovative power of space design; and, as a result, many projects, even amongst the most recent, lack the aesthetic quality of truly valuable architecture. Despite the fact that most of these companies, including almost all Fast Food chains, run their own architectural department, the innovative intensity associated

[27]Doing the "Wright" Thing (2003) In the Business Week/Architectural Record Awards special report. *BusinessWeek*, November 3rd.

with the usual lodging and restaurant facilities doesn't rank high enough. Restaurant units being frequently located in mall, looking at commercial centers and facilities associated with successful retailing enterprises constitute the natural prolongation of this first round of observations.

Creative Architecture for Success in Retailing

Retail has proved a major part of the life of millions of inhabitants of North America, Europe, and increasingly of Asia and Latin America. A dominant service activity in modern societies, retail, takes multiple building forms: Individual or chain stores, downtown department stores, or large commercial centers. More and more complex conglomerations extend beyond retail and combine entertainment in different forms. In modern times, the quality of the spatial representation of the retail physical support has obviously not been considered an important matter, as demonstrated by so many situations where the gas station along with strip malls and stores, lined up along the highway; stand unrelated in isolation on suburban America and Europe. On both sides of the Atlantic, multiple reports have condemned the negative effect on the landscapes of town entrance of the out-of-control proliferation of retailing facilities, though reports with very little apparent near-term impact. The economic growth since the 60s combined with the spectacular development of the supermarket formula, as well as the relatively loose regulation regarding the building of commercial spaces, except in Germany and to a certain extent Switzerland and Austria, have proved stronger forces than any other aesthetic consideration.

Architecture innovation definitely played a crucial role in the deep revolution that profoundly reshaped both the cities' downtowns and the retail trade and service sector. In a contribution on the architecture for the retail trade, Eberhard H. Zeidler identified three major causes that have wrought the current state of retail and have significantly influenced its future course: (1) The influence of transportation, particularly the automobile, (2) The discovery of emotion as a factor in the design of retail facilities, and (3) The necessary integration of retail into a fragile environment, which is now the concern of many.[28] For Zeidler, the 19th century "passage" fed on urban density and strengthened the urban texture, even after the introduction of trains and streetcars.

[28]Foreword by EH Zeidler (1996) In W Hocquel and JD Gartrell. *Architecture for the Retail Trade*. Basel: Birkhâuser.

Over the 20th century the automobile has proved to be the most crucial factor that changed shopping patterns. J. B. Jackson, a specialist of landscape history, writes: "The automobile, especially the commercial automobile, the truck, the pickup, the van and minivan and jeep has been most effective in introducing a different spatial order. For what those vehicles contain (and distribute) are not only new attitudes toward work, new uses of time and space, new and more direct contacts with customers and consumers, but also new techniques of problem solving".[29]

Organizational Zoom...

Space Design for Retail Development in North America

In North America, the relative freedom to build has dramatically altered the urban texture, and the extents of the resulting environmental damage are only now being sensed. The shopping centre created by Victor Gruen in the Detroit East Mall was probably the first experiment to transplant the city into what was previously a virtual no-man's land only reachable by car, opening the way to the multiplication of commercial malls.

Earlier, in 1916, Howard Van Doren Shaw designed Market Square, a combination of small stores and a Marshall Field department store for the Chicago garden suburb of Lake Forest.[30] The "historical" American city deteriorated even more due to this powerful trend, though the phenomenon was not immediately perceived as potentially negative since most citizens might have disliked their cities anyway due to the very specific history of urban development in the New World.[31]

Development of shopping centers in the United States slowed down due to the depression and World War II. In 1956, Victor Gruen, an admirer of the glass-roofed, 19th century gallerias of Milan and Naples, created in Edina, near Minneapolis, the first indoors, air-conditioned public walking areas for the Southdale shopping center.[32] Then, from 1960–1970, more than 8,000 new centers, mostly indoor shopping malls, opened in the United States, and by 1995 more than 40,000 operating shopping centers were registered, including about 2,000 so-called regional malls.[33]

[29]JB Jackson (1994). *A Sense of Place, a Sense of Time*. New Haven: Yale University Press.
[30]W Rybczynski (1996) *City Life*. New York: Touchstone, Simon & Schuster.
[31]*ibid.*
[32]V Gruen (1964) *The Heart of Our Cities: The Urban Crisis, Diagnosis and Cure*. New York: Simon & Schuster.
[33]International Council of Shopping Centers statistic.

Designed in 1970, the Eaton Centre in Toronto appears to be the first expression of an attempt aiming at integrating a "regional" shopping mall flawlessly into the fabric of the city.[34] The Eaton Centre building, with about 300,000 square meters of retail space, also took into account emotional elements in its architecture, displaying an innovative galleria able to create a sense of place. It ended up as one of the most successful commercial centers in North America.

Apart from a few number of other exceptional achievements, most of the retail developments that emerged during the period from the 60s toward the end of the 80s are not credited for having improved the aesthetic quality of the urban environment, neither to have fostered architectural innovation. Witold Rybczynski writes, "The architecture of most contemporary shopping malls is downright modest compared with the high-flown, extravagant interiors of the Victorian shopping bazaars and department stores which celebrated shopping and consumerism on a scale unrivalled before and since".[35]

For Giorgio Armani, the founder of the famous clothing company, "a boutique is a superbly effective means of communicating, through which you establish your trademark, your image; with the public (...) you reveal a moment of truth. That "truth" lies in how closely you've integrated the product's quality with its presentation".[36] Reinforcing the point, Grant Camden Kirkpatrick, an architect who, in the mid-90s, published in a noted book a remarkable collection of reports on innovative shop design,[37] adds: "the relationship between a store's image and its product is the major challenge to its success". Of course, one has to realize that working on a new concept of store usually involves interior design much more than traditional architecture. Both fields certainly belong to the broader subject of corporate space design and refer to industrial design as well. But, working on a store design produces a more ephemeral effect on the visual environment than building a new structure. Because of this ephemeral characteristic and of the usually limited level of required financial investment, the perceived level of risk seems lower, allowing for potentially more creativity. On the other hand, the impact of the generated new image and atmosphere on customer willingness to do business with the company appears to be taken more seriously than ever before. And many chain owners defend that it's easier

[34] Zeidler (1996). *Op. Cit.*
[35] W Rybczynski (1996). *Op. Cit.*
[36] G Armani (1994) Foreword. In GC Kirkpatrick. *Shops and Boutiques*. New York: PBC International, Inc.
[37] G Armani (1994). *Op. Cit.*

to touch customer through appropriate shop interior design than through grand architecture.

Most of the time, the store will use an already standing building which history might have nothing to do with retailing. And sometimes, the most valuable architectural impact of the effort to redeploy a new retail space precisely results from the conversion and revitalization of classic or industrial pre-existing structures. Anthropologie, a 6,500 square foot specialty apparel, and furniture retail space in Wayne Pennsylvania, as well as the Urban Outfitters apparel and accessories outlet in Santa Monica, California, display examples of successful conversions of historic buildings: A 1930 warehouse in Wayne, and a 1927 industrial structure in Santa Monica. Both places designed by Sue Otto and Ron Pompei, use a distressed and "broken" palette of materials such as natural, aged wood, and metals with oxidized patinas, simulating the familiar haven of the targeted consumer. Actually, a full commercial center designed on similar but more extremely expressed principles opened its doors in 1994 in Orange County, California. The place, officially called the Lab, and unofficially, the anti-mall, was displaying plenty of attitude, decay, and patchouli. Its builder, Shaheen Sadeghi, a former surf-wear executive, was targeted with this phrase "the lost generation of mall rats. At age 18 to 30, they are too cool for the food courts and too rumpled to appreciate Chanel or Nordstrom".[38] By 2004 in Europe, this given style was not yet been followed by a significant number of retailers, who were usually more comfortable with a much sober expression of modernity.

For Kirkpatrick, developers of places like Wayne's Anthropologie and Santa Monica's Urban Outfitters concentrated on "Store Imagery", where the projects tended to directly adapt the image of the target customer and merchandise to the architectural space. Very different stores would also belong to this category, such as the alluring and quiet Bergdorf Goodman Men's apparel 40,000 square feet luxury store in New York,[39] which became the trademark for its products and name, or the Ralph Lauren unique town house on Madison Avenue in the same city. Also, the Greco-Roman styled Ceasar's World,[40] a major gift and souvenir shop in Las Vegas,

[38]S Drucker (1994) The anti-mall. *The New York Times Magazine*, October 9th.
[39]Bergdorf Goodman Company, architect/designer of the NewYork Men's store: J.T.Nakaoka Associates Architects, Los Angeles, California.
[40]Caesar's World Company, architect/designer for the Las Vegas Gift shop: TSL: Merchant Design Group, Los Angeles, California.

Nevada, directly emulated the products in order to elicit memories and emotions. A second category of innovative design for store tends to respond to the challenge of promoting sales environment with sensitivity to economic parameters. For Kirkpatrick, the challenge took many forms,[41] including retail entrepreneurs who are exploring new ventures with limited resources, like the pH Neutro perfume shop in Florence, Italy for instance,[42] stores that seek the creative expression of inexpensive materials to suggest the value of their products, such as the Levi's Only casual apparel in Ontario, Canada,[43] and retail owners who "desire the illusion of champagne space within a beer budget", the St. Mark's Bookshop in New York offering an attractive case.[44]

Finally, one may recognize a third category of projects that probably corresponds to the most creative design efforts, namely the "Unconventional Venues". A conversion of a Beaux-Arts style bank building, the 12,000 square feet Emporio Armani apparel store in San Francisco includes a central bar as the focal point of the ground floor, while the mezzanine level houses the dining area of a dedicated restaurant.[45] In this case, the creative value comes from the fact that the retailing space integrates special features, such as bar, coffee point, or restaurant, as focal points for their sales areas to creatively display new merchandise. The Sara Sturgeon apparel outlet in New York was thought about as an interpretation of some sort of medieval fortress allowing the architect, Raffi Balouzian, to compose a thematic approach of the retail store and showroom area. More and more retailers explore creative means to combine showrooms, and spaces for the display of fashion merchandise, with shops. For promoting advanced products, such as hifi systems or high end furniture, some retailers develop areas where merchandise is used to relay a sense of comfort of home with style, as seen at many of the retail franchises of the Danish Bang & Olufsen audio–video manufacturer.

---

[41]Kirkpatrick (1994). *Op. Cit.*

[42]pHNeutroperfumr Company, architect/designer for the Florence, Italy, shop: Jensen & Macy Architects, San Francisco, California.

[43]Keith Kovar owner, Levi Strauss & Co. franchise in Canada; architect/designer: The International Design Group, New York, New York.

[44]St. Mark's Bookshop, Bob Contant, and Terry McCoy associates; architect/designer: Zivkovic associates Architects P.C., New York, New York.

[45]Georgio Armani Corporation; architect/designer of the San Francisco store: Thane Roberts, Santa Monica, California.

New in the debate in early 2000s, a handful of manufacturers of electronics products were following this trend toward the building of sophisticated retail space. In May 2001, Apple Computer Inc. announced its intention to launch a series of own shops. Following Apple, Sony Electronics Inc., and the PDA maker palmOne especially invested in creatively designed new shops aiming at displaying their most advanced products and above all at communicating on the philosophy underlying their strategies. However, most of the manufacturers, ranging from Gateway Inc. to Dell Inc., who had opened stores and kiosks quickly decided to shut them down. Justifying the opening of their 11 stores in malls, palmOne, which hands out rebate slips to people who visited its stores, found that 31% later bought palm products at other places or online.[46]

Organizational Zoom ...

How Every Square Foot of an Apple Store is Designed to Make You Spend More Money

The Apple Store is an incredible retail success. If you own a piece of Apple tech, there is a good chance you have visited one of their retail locations. Every detail of each Apple store has been designed and planned out to influence the customer into spending more money as they shop and browse (see Figure 8.1). In just one quarter of 2013, nearly 100 million people shopped at the company's 424 global stores.[47] According to the San Jose Mercury News, Apple stores generate more revenue per square foot than any other US retailer.[48]

One of the major reasons that Apple stores are so successful over the years is the spectacular design details that are put into each of the stores. For example, take its flagship store established in New York's Fifth Avenue. The wide open store, with a large skylight and a central staircase that appears to float up to a second story, is spare, with a small number of Apple computers and iPods displayed as objects of desire.[49] Dominant colors include white and light grey, and glass and aluminum constitute the most visible materials mobilized for this reconstruction signed by Californian architect Bohlin Cywinski Jackson.

---

[46] C Edwars (2004) Boutiques for flogging the brand. *BusinessWeek*, May 24th.
[47] http://www.businessinsider.sg/most-beautiful-apple-stores-2014-5/?op=1#.VGmTalOUdjE.
[48] http://www.mercurynews.com/business/ci_25822327/apple-store-geniuses-rare-peek-behind-curtain.
[49] Where apples really shine (2003) In the BusinessWeek/Architectural Record Awards special report, *BusinessWeek*, *Op. Cit.*

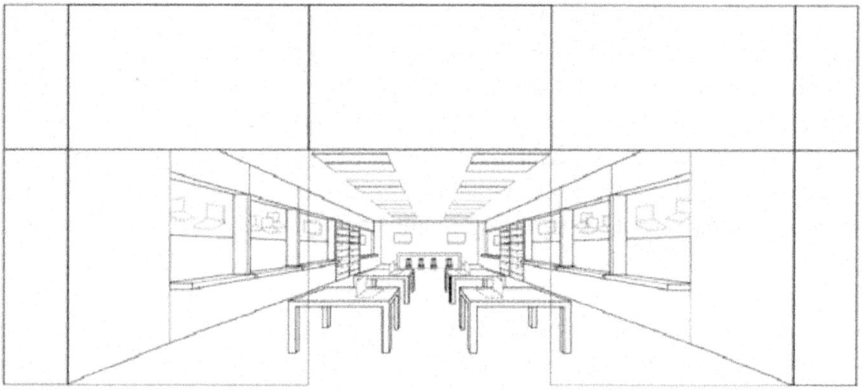

Figure 8.1: Apple has been granted a Trademark for the Glass-Fronted, Rectangular-Patterned Design of its Apple Store.

Source: Image of US Patent and Trademark Office.

> Inside, all of the stores look pretty much the same, but each one offers its unique look from the outside. In fact, Apple trademarked its store layout in 2013. The trademark consists of two designs, one in color and one in black and white. Both present the typical Apple Store layout, with wide tables in the middle and benches around the sides where customers can try out the latest products, as well as the Genius Bar help desk at the back. In the New York Fifth Avenue store, Customers are supposed to interact with the products by themselves and should enjoy a "party" atmosphere. Room for children to play with software is provisioned on the upper floor, which is adjacent to a 40-seat conference room. As a widely advertised about, the store is supposed to represent one of the company's key illustrations of its CEO's vision.
>
> The co-founder and then-CEO Steve Jobs believed that the Apple retail program needed to fundamentally change its relationship with the customer, and provide more control over the presentation of Apple products and the Apple brand message. As a matter of fact, architects, graphic designers, product developers, and merchandising people have been grouped together under the active leadership of Jobs to create a space that projects the philosophy of the company.

Other industries also rediscovered the value of investing in the design of innovative "shopping" areas. That proves to be especially visible in the automotive industry, in which the initiative was left to dealership almost from the origin of the activity. Daimler-Benz, which maintains in Germany its ownership over numerous points of sales, naturally showed the way in defining more than usual sophisticated and standardized outlets. An original project, called the A-Class Box,[50] was initiated for the commercialization of the lower range vehicle, quickly followed by a more ambitious and creative project for the Smart Car brand. On the famous Champs Elysées Avenue in Paris, the Renault building, formerly known as the Renault drugstore, has long established itself as a prime spot with its trendy bar and modern-style restaurant, while most of its competitors' outlets in 2004, with the exception of Toyota, still display traditionally designed dedicated show rooms. BMW also appear to spend energy in defining innovative concepts for the design of its new showrooms and aims closely at assisting its dealerships in the design of their own commercial areas.

Petrol groups have always paid attention in the design of their networks of gas stations, although not necessarily going for truly innovative solutions.[51] Some of them however, like British Petroleum for instance, tried to improve the aesthetic quality of their points of sales as showed in interesting experiments in France and Switzerland in the early 21st century. In itself, the study of the mobilization of the space design and architectural resources by the key players of the large automotive industry would deserve an entire series of thick books. So, we will not elaborate on this topic here. Other commercial activities, even the one based on pure intangible value creation, like the financial services, the use of which do not show much presence in regular shopping centers and malls, have eventually massively invested into new formulas and creative physical representation. But banks have a long history of architectural expression in Cities and through all forms of representations: Buildings, brochures, checkbooks, banknotes, and websites.

The Future of the Office Building

In the 19th and 20th century when free market and triumphant capitalism were dominant, banks and insurance companies were competing for erecting the highest and more dramatic buildings ever. And very often, the medium constitutes the message. In North American city centers of New York or Chicago, the early skyscrapers of the 1920s and 1930s offer panoply of

[50]The A-class box: A modular displaying system designed in 1996 by Tobias Wulf + Partner, architects in Stuttgart.

[51]A Jakle and KA Sculle (1994) *The Gas Station in America*. Baltimore and London: The Johns Hopkins University Press.

heavy handed symbolism, however dominated by the great manufacturers of the time. Bank of America in Chicago with its magnificent lobby, a 1923 design by Jules Guerin, illustrates the "grand-architecture" applied to the New World financial institutions. Building activity in the US naturally slew down just after the winter of 1929–1930, due to the financial crisis, apart from the great exceptions such as the Rockefeller Center, almost the only big architectural project during the early 30s in New York.[52]

And then, another dominant style emerged in corporate design. Escaping from Germany during the 30s, Mies van der Rohe introduced the new International Style, sober design with glass and visible iron made dark frames, which drastically influenced architecture for the following decades. Later on, variants emerged, such as the 100 stories John Hancock Center in Chicago, a prototype for a tall office building as a city within a city. In Europe also, similar approaches influenced the design of banking facilities. Goldman Sachs European headquarters in London, at Peterborough Court-Fleet Street, is a vocal example of power-style. The "Muscular" building speaks of power, of deal-flow, of secret knowledge.[53] By mid 90s, opulence is, once again, considered obsolete, at least in Europe and North America. In fact, whatever their business purpose, the tallest skyscraper most recently erected are now located in Asia and the Middle East, such as the 2010 Dubai Burj Khalifa culminating at 824 meters, the 2011 Tokyo Skytree at 634 meters, the Shanghai Tower World Financial Center at 632 meters, or the Makkal Royal Clock Tower at 601 meters. Selection also include the 509 meters high Taipei 101 in Taiwan, and the 1998 twin Petronas Towers in Kuala Lumpur, both 452 meters high. Since financial services and other office-based businesses in the West believe that they have entered a permanent change mode in the late 90s with modern architecture favoring adaptability and utility.[54] Adaptability and functionality also suit real estate management objectives since they give buildings higher resale potential. Naturally, the style, now undemanding and modest, paid a tribute to the new trends and projection of identity could no longer be strongly conveyed; thanks to architecture. Most of the interior of the banking buildings between 1995 and 2000, for instance,

[52]D Okrent (2003) *Great Fortune*: *The Epic of the Rockefeller Center*. New York: Viking.
[53]S Cooke (1996) Towards a machine for working. *Euromoney*, August.
[54]*ibid.*

relied on standardized organization and commonplace features for trading, settlement, and corporate finance. J.P. Morgan's London office and Barclays headquarters, also in the City, presented good examples of modest and functional designs, which despite the low key orientation preserved a sense of elegance and quality. Also, some developers were realizing that users of office facilities have become more demanding and companies, especially in the Benelux and Northern Europe, have adopted innovative approaches in the design of office space. Frank Duffy and Alex Henney wrote in an influential book on the modern evolution of city: "The great need is for an alternative image of the office to that of a money-multiplying, inhuman, non-life-enhancing tower".[55]

Years ahead of its time when it was finished in the 1980s, ING bank headquarters in Amsterdam is noted by Duffy and Henney as an excellent example of user-friendly building. On top of providing an exceptional two-level Japanese-style garden for the employees to relax in, the corporate village of ten low towers whose walls slope in multiple directions offers restaurants, snack bars, post room, library, film theatre, and conference rooms, all reachable through an indoor walkway. As opposed to a common North American practice in modern offices, no worker sits more than six meters from an opening window. Interesting initiatives also include the erection of the Bank of Luxembourg headquarters in 1995 on a design by Brescia and Spare from Arquitectonica, considered the most creative, even provocative architectural firm in Florida since the late 70s. Unfortunately, due to endless internal discussions, the treatment of the facades look rather commonplace, while amazing space organization awakes the interior.[56]Also, in Geneva, Mario Botta had been hired by Banque Bruxelles-Lambert to build its medieval-castle-type Swiss headquarters.[57] Already known for its architectural audacity, the bank worked during the 50s with American architect Gordon Bunshaft, from the famous Skidmore, Owings & Merril firm of Chicago, to build its central tower in Brussels. Among multiple cases in North America, the Federal Credit Union in Oklahoma, a relatively low-key structure designed by Elliot & Foltz, display wooden frame and stone walls, with glass encapsulated meeting rooms opening on interior gardens

[55]F Duffy and A Henney (1989) *The Changing City*. London: Bulstrode Press.
[56]F Eldemann (1995) Des architectes de florideconstruisent la nouvelle banque de Luxembourg. *Le Monde*, May 12th.
[57]M Champenois (1996) La banqueetl'architecte. *Le Monde*, April 29th.

with luxuriant vegetation. But this time, the building in Geneva stays on the sober mode.

Organizational Zoom ...

SociétéGénérale Tower, Paris, La Défense, Quartier Valmy

In contrast to the tendency toward less formalism and more user-friendly atmosphere, a few large financial institutions keep on racing for the most dramatic architectural statement.

Founded in 1864 in Paris, SociétéGénérale (SocGen), the second largest banking corporation company in France, probably belongs to this category. Already in 1871, the opulent new headquarters by Hittorf stands proudly on what soon became the Boulevard Haussmann, named after the famous urban planner who entirely remodeled Paris under Napoléon, the third. In 1987, studies unveil the necessity to relocate the bank's headquarters and to group together the many teams spread all over the Parisian metropolitan area.[58] A first proposal for installation in La Défense, the business complex west outside of Paris, was refused in 1988 by public authorities, but finally obtained approval in 1991 after many alterations of the initial project. La Défense already housed 14 of the top 20 French corporations when SocGen began erecting its new building there. Designed by architects Andrault-Parat and Ayoub, and completed by June 1995, the 167 meters high company's twin tower offers 200,000 square-meters of floor space, excluding basic structure. With an area per standard floor of 1,300 square meters, it accommodates 6,000 employees, 17,500 square meters of trading rooms, and 1,400 parking places. With a US $1.1 billion investments in 1995, the project was partially financed by the sale of an already existing tower in the same district, and by the release of 63,000 square meters of office space in Paris downtown, including the 8,000 square meters of the historic headquarters.[59]

A certain level of innovation took place in the way technology was implemented. For instance, each employee was equipped with a sophisticated multi-media communication system, as well as a command allowing for individualized lighting, temperature, and fan. The trading rooms are equipped with sound reducers that significantly cool down the atmosphere of these usually over heated playgrounds.[60] The

[58] La Tribune Desfossés (1995) La SociétéGénérale se dote d'un nouveau siege. *High-Tech*, October 26th.
[59] Le Figaro (1995) Unnouveau siège pour la SociétéGénérale. October 26th.
[60] A Demerlé-Gof (1995) Les tours de la Générale. *Le Moniteur*, October 27th.

"free-cooling" air-conditioning system starts working when external temperature reaches 13°C for energy saving purpose and water from condensation is reused for hot water batteries.[61] Finally, a computer-aided maintenance control system analyzes operating cost on a regular basis, an important function since then. According to Duffy and Henney, the 60 year cost of a corporate building is dominated by space plan, redone seven to 10 times, and skin services, redone two to three times.[62] Almost ten years after its completion, opinions differ regarding the adequacy of the choice for the huge and dramatically formal facility for efficiently housing the multiple and very diverse banking activities of SocGen.

Recognizing the architecture's ability to relevantly address corporate problems, increase workers productivity, and boost performance, companies started to explore new paradigms. In the late 90s, architect HOK turned an old factory in the US into flexible, low-cost space for Northern Telecom,[63] paving the way for a series of new experiments. The clean, well-lit Astra France HQ space was noted in 1999 for its emphasis on transparency.[64] In 2012, In the gleaming five-story research and marketing building of Finnish mobile Nokia in Sunnyvale, California, fewer than 10 out of the 500 employees had a permanent workplace. Employees work over cappuccinos and tasty sandwiches from each floor's café. Two wellness rooms, with showers and napping areas, help traveling executives decompress between challenging meetings.[65] Clusters of arcade games and a case of obsolete Nokia models are constant reminder of the inexorable need to innovate or perish. However many studies are unveiling the need for many workers to have access to privacy that is usually not provided in open spaces.[66] On the other hand, it is said that a musical-chairs approach to an office may improve efficiency, as illustrated with the Gerson Lehrman advertising company or at the 6,500 square-meter Google's office in Manhattan, where employees are equipped with laptops and headsets and can choose from a variety of unassigned workplaces.[67] More offices are offering their staff pantries and cafes with

[61] Le Moniteur (1995) Climatisation modulable pour le tour de la SociétéGénérale. April 4th.
[62] F Duffy and A Henney (1989). *Op.Cit.*
[63] B Nussbaum (1997) Blueprints for business. *BusinessWeek*, November 3rd.
[64] BusinessWeek (1999) Good-looking hardworking. November 1st.
[65] A Braithwaite (2012) Helsinki in the valley. *Bloomberg BusinessWeek*, April 30th.
[66] D Bennett (2013) Why we can't get anything done in an open-plan office. *Bloomberg BusinessWeek*, October 10th.
[67] B Lanks (2014) Don't get too cozy. *Bloomberg BusinessWeek*, September 22nd.

snacks and drinks in casual settings, such as Barclays in London, Viacom or Bloomberg.[68] Advertising agencies naturally show the most adventurous initiatives in the matter of their own space organization. Fnuky's headquarters in Adelaide, for instance, picked the sumptuous Victorian bordello style as their design theme,[69] while Bailey Brand Consulting in Plymouth meeting, Pa., parked a former London double-decker transit bus in the lobby of its office loft-building, while the data-storage equipment is housed in a climate controlled classic Airstream trailer nearby.[70] Companies from the Internet economy are also following the trends as expressed in the design of Airbnb's San Francisco headquarters,[71] among others, which features replicas of some of the most creative rooms on the site.

Organizational Zoom ...

Innovative design at UBS in Switzerland

Swiss Bank Corporation, before its merger with Credit Suisse, developed a standardized customer exchange area articulated around three zones: A self-service side close to the entrance, a central space devoted to advice and sophisticated services, and finally a still maintained traditional glass-counter side. For the user, the attended advantages of this design include an immediate viewing of the three zones, the possibility of choice between fast-track machine-based transactions or personalized treatment, as well as the extension of opening hours of the self-services area.

At the Union de Banques Suisses regional head-branch in Lausanne, based on similar principles, a more elaborated design was implemented in 1994 and was finally been abandoned when the branch was relocated in 2010. Opening on the main local plaza, the first slide of the large and lightly lobby was packed with advanced automated teller machines which communicated to a second area housing the "Première Loge".

An innovation in the Swiss banking landscape, "Première Loge" service offered, from 9.00 am to 4.30 pm, booking for theatres, opera, and music shows all over the country. Then, customers switch to the main business zone, starting with an information desk and waiting space furnished with Le Corbusier famous LC leather sofas. Traditional counters gave space for customers' adviser spots. For investment and private counseling, a set of quiet lounge suites stood in the back of the building.

[68]K Goh (2012) Grab a bite and chill in the office. *The Sunday Times*, May 27th.
[69]M Siegel (2012) Creative spirits. *Bloomberg BusinessWeek*, March 26th.
[70]W Spielgelman (2012) Minds on the move. *Bloomberg BuinessWeek*, March 5th.
[71]J Hempel (2012) Airbnb: More than a place to crash. *Fortune*, May 21st.

Still in operation by the middle of the 2000s, this lay out has proved satisfying to both the customers and the bank employees. Initial self-service packages usually included cash machines, currency changers with offline features, banknote and coins converters, and an emergency telephone, directly linked to the closest police station, fire department, hospital, and UBS 24 hour switchboard. Then, the basic set of equipment has been considerably extended in the decade following the implementation of the new branch concepts in banking all over the country. Fully-automated branches, offering advanced functions, became a commonplace feature in Swiss cities main streets and malls as early as 2000. Similar patterns have influenced the evolution of the "physical" landscape of banking in Europe and North America.

New Technologies and the Design of Virtual Space

Importance of corporate design might paradoxically be enhanced by the very fact that the necessity of being related to building to deliver services is now questioned by the development of new technologies in many sectors.[72] Education, for example, has traditionally been provided in classrooms. Schools and universities mobilize significant facilities that very often constitute major urban landmarks. While education seems to become a dramatically growing for-profit service activity, e-learning through massive online open courses (MOOCs) or other initiatives is exploding and fueling the development of organizations offering distance learning.

With 175,000 salespeople and service agents in more than 7,500 dealerships at its peak, General Motors Corp. has spent a fortune bringing employees to hotel rooms and classrooms for training. Using interactive distance learning technology now being installed at every dealer, IDL has let employees view a live course beamed in by satellite, asking questions to the instructor, without leaving their dealerships. Even the US Army has jumped on the bandwagon offering more than 1,000 different courses in information technology over the internet by SmartForce in the late 90s. E-learning isn't limited to tech training. Shoney's chain of restaurants has begun training waiters, cooks, and other employees using a novel satellite-delivered computer program to teach recruits such basics as how to clock in for work or to take order.[73] In year 2000,

---

[72]H Mathe and TF Dagi (1996) Managing technology for the globalization of service operations. *International Journal of Technology Management* 12(5–6). pp. 577–607.
[73]W Symonds (2000) Education. *BusinessWeek*, January 10th.

International Data Corp., predicted that e-learning would explode to a $7.1 billion business by 2002. As a matter of fact, their forecasting proved pessimistic.

However, classroom-based courses are not going away; but institutions and training firms that relate to mediocre facilities while still focusing on traditional classroom-type services will see their market shrinking. It is critical that more exceptional customer exposure to service facilities do take place. Actually, development of telecommunication services and internet services generate the multiplication of new retail shops. NTT DoCoMo has to implement a very dense network of hundreds of outlets across Japan to distribute mobile phones and subscriptions. US-based Lycos Web portal announced in the end of 1999 a joint venture with Singapore Telecommunications to set up customized versions of Lycos in 10 Asian cities.[74] As a tangible result of this alliance, Lycos Asia's physical artifacts already pumped out in the streets of Singapore and became a remarkable component of the urban landscape.

Sometimes, the connection with a tangible facility appears only as a suggestion since the customer interacts with the service provider on the Web only, such as in e-banking. If the internet has proved one of the very dominant vehicles for innovation in the financial industry, the reference to real buildings is still very much present in the way online banks communicate with their customers. A content analysis of the top 10 US banks, which was realized in the late 90s by Jonathan E. Schroeder, from the Royal Institute of Technology in Stockholm, revealed that over 50% customers use some form of architectural image on their websites. Furthermore, several of these financial companies rely heavily on buildings, headquarters, and the classical motifs for web-based imagery.[75] So, whatever the service sector, firms have to consider architecture and space management as a key strategic resource. Then, innovating in the way service facilities are designed and managed represents a stimulating and permanent objective for any company willing to consolidate its competitive situation.

[74]I Khnii and S Baker (2000) Amazing DoCoMo. *Business Week*, January 17th.

[75]JE Schroeder (2003) Building brands: Architectural expression in the electronic age. In LM Scott and R Batra. *Persuasive Imagery: A Consumer Response Perspective.* Mahwah, NJ: Lawrence Erlbaum Associates, Publishers, pp. 349–383.

VALUING SPACES AND DESIGN PROJECTS: PROCESS AND METHODS

The lack of formalized methods to appraise the possible economical impact of architectural change for the investing company contributes to significantly limit the willingness of company for experimenting really creative designs. Appropriate use of articulated methods certainly proves necessary for any service company willing to objectively make key design decisions. Our objective, at this point of the discussion, is not to propose a comprehensive toolbox allowing for a full appraisal of the performance of service architecture. We simply would like to introduce a three level approach mobilizing available principles in view of contributing to a first evaluation of design projects or existing service facilities. This approach aims at contributing to bring answers regarding the following three major questions:

- How to adequately define design tasks in order to maximize the chance of success for any future service facility?
- How service buildings can be designed and used to convey the expected image to beneficiaries, employees, shareholders, and the social environment?
- How service facilities work? May the location, spatial layout, and space use be considered efficiently enough? Are the contact networks generated by the spatial organization in line with the company's objectives and useful for the people involved?

In order to efficiently define the design task, one may follow on the path suggested by Thomas Walton in Architecture and the Corporation.[76] The author believes in improving the understanding of the relationship between architects and executives by postulating "laws" to account for their respective expectations. Four "maxims" summarize the conceptual basis for using design effectively as a business resource: (1) Integrate multiple perspectives; (2) Emphasize casting; (3) Sift for relevant facts; (4) Manage a pluralistic decision-making process (see Table 8.1). Then, the question of the "imageability" of the productive space should be carefully studied. Kevin Lynch, in The Image of the City, defines the concept of "imageability" as the quality in a physical object that gives it a high probability of evoking a strong image in any given observer. It is that shape, color, or arrangement which facilitates the making of vividly identified, powerfully structured,

[76]T Walton (1988). *Op. Cit.*

Table 8.1 Maxims to Maximize Results in Design.

Four-Step Program	Corporate Tasks
A. MULTIPLE PERSPECTIVES	
• The Managerial Perspective	— Define Objective and Priorities
	— Establish Standards of Quality
	— Articulate Corporate Value
• The Employee Perspective	— Seek Employee
• The Architectural Perspective	— Look for a Compatible Marriage with the Architect
B. CASTING	
• Multitude of Actions and Roles	— Appoint the in-house players
	— Hire outside managers
	— Select the design team
	— Work with key public actors
• Assembling the Cast	— Determine the Mix of In-House vs. Outside Services
	— Stress the CCC Syndrome: Commitment, Communication, and Competence
C. SIFT FOR RELEVANT FACTS	
• Design Research	— Turn facts into Options
• Critical Topics	— Study Technology, Layout, and Image
D. MANAGE PLURALISTIC DECISION-MAKING	
• Elements in Design	— Emphasize Goals, Not Positions
	— Develop Consensus
	— Integrate Decision-Making and Culture

Source: Adapted from Thomas Walton "Architecture and the Corporation", 1988.

highly useful mental images of the environment.[77] An environmental image may be analyzed into three components: Identity, Structure, and Meaning. Identity induces recognition as a separable entity, while structure refers to the spatial or pattern relation of the object to the observer and to other objects. Meaning could also be defined as a relation, but not so easily influenced by physical manipulation.

Corporate identity programs, which are developed by big firms or communities such as cities or public services, emphasize the use of

[77]K Lynch (1960) *The Image of the City*. Cambridge: The MIT Press.

"imageability". As a brand management specialist, Wally Olins explains that Corporate Identity programs mark out turning points in a corporation's life because of their high visibility and impact.[78] Increasingly, corporate identity programs are concerned with expressing three separate but interrelated themes- coherence, symbolism and positioning, which go to the very core of a corporation. The organization wants its different parts to relate to each other in a clear and comprehensive way. Any organization usually expects that everyone who works for it can share the same driving spirit and then communicate this spirit; on top it also needs to differentiate itself and its services from those of its competitors.[79] However, corporate executives and their identity enhancing consultants seem to only begin to really acknowledge the power associated with the mobilization of architectural resource in order to increase corporate "imageability".

However, a few product-design consulting firms have entered the field of service and commercial space design. An interesting illustration, IDEO, a Palo Alto (California) company, has designed hundreds of products, especially laptop, PDAs and other high-tech devices, and won more design awards over the 90s than any other firm.[80] Now, IDEO, the 350-employees company, is transferring its ability to create consumer products into designing consumer experiences in services, from shopping and banking to health care and wireless communication.[81]

For instance, the largest health maintenance corporation in North America: Kaiser Permanente, hired IDEO in 2003 to help thinking the next-generation medical offices and hospital buildings they would have to put in place to cope with their long-range growth plan. First of all, IDEO's innovative five-step program approach called for the constitution of a special project team joining Kaiser's medical personnel and facilities managers with the designs firm architects, designers, social scientists, and other experts. Following phases of the project include an intense, idea-generating brainstorming session based on the data gathered by observing

[78]W Olins (1989) *Corporate Identity: Making Business Strategy Visible Through Design.* London: Thames & Hudson.
[79]W Olins and E Selame (2001) *The Corporate Audit: A Set of Objective Measurement Tools for your Company's Image and Reputation.* Cambridge: Cambridge Strategy Publications.
[80]IDEO began in 1991 as a merger between David Kelley Design, responsible for the creation of Apple Computer Inc.'s first mouse in 1982, and ID Two, which designed the first laptop computer also in 1982. Office-furniture manufacturer Steelcase Inc. owns a majority stake in IDEO.
[81]B Nussbaum (2004) The power of design. *BusinessWeek*, May 17th.

people. Rapid prototyping of product but also of services and spaces may then help visualizing possible solutions and speed up decision-making and innovation. Engaging clients actively in the process of narrowing the choices to a few possibilities proves determinant to the success of the project. Finally implementation brings the firm's capabilities in engineering, design, and social science to bear when actually creating a product or service (see Table 8.2).

At Kaiser health care organization, observation of patients have allowed for series of insights, like revealing that patients and families feel quickly annoyed since checking-in is considered a nightmare and waiting rooms uncomfortable. Also, patients have difficulties with examination rooms because they often wait for up to 20 minutes half-naked, surrounded by threatening needles. "IDEO and Kaiser concluded that the patient experience can be awful even when people leave treated and cured".[82] After a couple of months since the start of the project, the health care group realized its long-range growth plan didn't necessarily require erecting numerous expensive new facilities. Seeking medical care is, after all, a social experience. So Kaiser had to implement simple recommendations to offer more comfortable and user-friendly surroundings and more efficient layouts, at relatively low capital expenditure. Apparently, many service companies still don't really know how to best connect with their consumers. It's why the like of IDEO's services have been in growing demand.

In order to answer the interrogation concerning how service facilities work, numerous directions may be explored. We have decided to concentrate on one aspect of this rich problem, among many others that cannot be developed here. The objective of the analysis is to establish a description of underlying spatial structure in order to better understand two kinds of relations: Those amongst the inhabitants of the systems, i.e., the customer, service providers, etc., and those between inhabitants and strangers. Introduced in the early 80s, the "space syntax" theory has developed into an extensive research program aiming at evaluating spatial configuration and functioning patterns created through the design of buildings and cities.[83] Shape recognition techniques transform the plan into a mathematical

---

[82]Nussbaum, B. *ibid.*

[83]B Hillier (1996) *Space is the Machine.* Cambridge: Cambridge University Press. See also: Young Ook Kim, Hoon Tae Park, Kyung WookSeo (eds.) (2013) *Proceeding of the Ninth International Space Syntax Symposium.* Seoul, South Korea: Sejong University Press.

Table 8.2 IDEO's Process of Designing a Better Consumer Experience.

Five-Step Program	Tasks and Techniques
1. OBSERVATION	
IDEO's experts team up with the clients to understand the consumer experience	— Shadowing *Observing people using products or experiencing services* — Behavioral Mapping *Photographing people within a given space over two or three days* — Consumer Journey *Keeping track of the interactions a consumer has with product, service or space* — Camera Journals *Asking consumers to keep visual diaries of their experience* — Extreme User Interviews *Evaluating experience of people who know-well or know-nothing about the service* — Storytelling *Prompting people to tell personal stories about their consumer experiences* — Unfocus Groups *Interviewing a diverse group of people to explore ideas about a given service*
2. BRAINSTORMING	
Intense, idea-generating session analysing data gathered by observing people	— Defer judgment *Don't dismiss any ideas* — Build on the Ideas of Offers *No "buts"*, only *"ands"* — Encourage wild ideas *Embrace the most out-of-the-box notions* — Go for Quantity *Aim for as many new ideas as possible* — Be visual *Use color markers to write on big Post-its that are put on a wall* — Stay focused on the topics *Always keep the discussion on target* — One conversation at a time *No interrupting, no dismissing, no disrespect, no rudeness*
3. RAPID PROTOTYPING	
Mocking up working models helps everyone visualize possible solutions and speeds up decision-making and innovation	— Mock up every thing *Models can be created not only of products but also of services and spaces* — Use videography *Make short movies to depict the consumer experience* — Go fast *Build mock-ups quickly and clearly*

(*Continued*)

Table 8.2 (*Continued*)

Five-Step Program	Tasks and Techniques
	— No Frills *Make prototypes that demonstrate a design idea without sweating over the details* — Create Scenarios *Show different ways to use a service and how various designs can meet their individual needs* — Bodystorm *Delineate different types of consumers and act out their roles*
4. REFINING	
At this stage, IDEO narrows down the choices to a few possibilities	— Brainstorm *in rapid fashion to weed out ideas and focus on the remaining best options* — Focus prototyping *on a few key ideas to arrive at on optimal solution to a problem* — Engage the client *actively in the process of narrowing the choices* — Be disciplined *and ruthless in making selections* — Focus *on the outcome of the process — reading the best possible solution* — Get agreement *from all stakeholders, especially including top-level executives*
5. IMPLEMENTATION	
Bring IDEO's capabilities in engineering, design, and social-science to bear when actually creating a product service	— Tap all Resources *Involve IDEO's diverse workforce from 40 countries to carryout the plans* — Manage appropriate workforce *Appropriate team include advanced degrees holders in different kind of engineering and associate subjects as well as experts in fields as diverse as fashion, communications, linguistics, ergonomics, art therapy, ethnology, medicine, and zoology.*

Source: Adapted from R Carra (2004) This is the IDEO way. *BusinessWeek*, May 17th.

network that can then be analyzed. On the basis of space syntax analysis, it has been proved possible in numerous cases to explain how inhabitants move. Nevertheless, no explicit representations of either motivations or individual cognition are actually mobilized.[84] The method has considerably evolved and improved over time. Nowadays, it seems commonly used for

[84]A Penn (2001) Space syntax and spatial cognition: or why the axial line? *In Proceedings of the Third International Space Syntax Symposium.* Atlanta: Georgia Institute of Technology.

explaining the success or failure of housing development or administrative buildings.[85] Given its predictive power, an extensive range of architectural and urban designs through the space syntax applications have also been done. The predictive quality of the space syntax analysis relies on the observation that it implicitly embeds the way people understand their environment and decide on movement behaviors.[86]

For instance, space syntax analysis revealed the underlying spatial orders of a failed commercial mall in Denver, Colorado. The approach allowed for the comparison between this recognized spatial structure with the one of another relatively similar, but successful, mall; its functional deficiencies were then identified.[87] Studies on the spatial types of museums and research laboratories have also mobilized space syntax techniques.[88] A recent investigation led to the conclusion that there is an underlying "genotypical" conflict in modern museums between the need to congregate people and the need to organize their movement. While achieving a study of service space syntax, a few generic tasks have to be performed in strict order, starting with a technical investigation and then being followed by a sociological analysis as explained by Space Syntax website, a consultancy established by the space syntax research team at University College London.[89]

Christian Norberg-Schulz asserts that human life is largely conditioned by the quality of its "existential space", which itself affects the image of the environmental structure. Major social entities and fundamental elements of the environment belong to such a structure. Indeed, the nature of the physical surrounding influences the quality of the interactions between people. Space design also affects customer expectations and choice[90]; and smells and sounds also play key roles.[91] Equally important, physical

---

[85]A Penn and P Turner (2000) A system and method for intelligent modelling of public spaces. UK Patent Office Application No. GO28477.

[86]Saif-ul-Haq (1999) Can space syntax predict cognition? *In Proceedings of the Second International Space Syntax Symposium.* Brazilia: Universidad de Brazil.

[87]M Brown (1999) Design and value: spatial form and the economic failure of a mall. *Journal of Real Estate Research* 17 (1–2). pp. 189–226.

[88]J Peponis and J Hedin (1982) The layout of theories in the Natural History Museum. *9H* No. 3. pp. 21–25.

[89]www.spacesyntax.com

[90]GA Churchill and C Surprenant (1982) An investigation into the determinants of customer satisfaction. *Journal of Marketing Research* (19), November. pp. 491–504.

[91]J. Baker, D. Grewal, A. Parasuraman (1994) The influence of store environment on quality inferences and store image. *Journal of the Academy of Marketing Science.* Vol. 22, Issue 4 pp. 328–339.

surroundings have a significant influence on employee motivation and effectiveness.[92] Customers take into account physical surrounding when deciding for a given transaction, but they also consider it while constructing their perception of satisfaction and need fulfillment during and after the encounter.[93] Increasingly, organizations have realized the importance of space design. In almost any sector, there would be a series of cases illustrating how successfully enterprises have made innovative use of space design in view of strengthening their competitive positions.

[92]E Sundstrom and I Altman (1989) Physical environments and work-group effectiveness. In: L.L. Cummings & B.M. Staw (Eds), *Research in Organizational Behavior.* Vol. 11. pp. 175–209.
[93]VA Zeithaml and MJ Bitner (1996) *Services Marketing.* New York: McGraw-Hill.

Chapter 9

IN SEARCH OF INSPIRING ENVIRONMENTS

Despite the unfavorable economic climate of the 2008–2014, there has been an increase in investment for innovation globally, especially in the vast majority of emerging economies where spending in R&D is resolutely ascendant.[1] Indeed, while the US participation in the global R&D effort reached 32.8% in 2012, the Asian participation increased from 34% in 2010 to 35.5% in 2011, eventually reaching 36.7% in 2012. Investments in Europe have remained stable, going from 24.8% in 2010 to 24.5% in 2011 and 24.1% in 2012.[2]

In Asia, China has been the largest contributor since 2010, with a total investment of 16.2% in 2012. In comparison, Japan's investment represents 11.8%, the equivalent of USD 148 billion for 2012. Although India is still far behind in terms of spending (2.8% in 2012), it has however developed highly efficient innovation mechanisms in specific fields that do not necessarily require significant investments. It is thus important to keep in mind that the impact of an R&D policy cannot be measured simply by the amounts of dollars allocated to it. As discussed in previous chapters, we believe the

[1]Some materials for this development are taken from H Mathe (2013) *Multinationals' R&D strategy in Asia: A driver for global innovation?* ESSEC-ISIS Cahiers Innovation & Society. Paris, December. ISBN 978-2-36456-087-1. Authors would like to thank Xavier Baraton, Group Vice-President, STMicroelectronics, François Luc, Vice-President R&D, Thales Solutions Asia, and Xavier Periera, Marketing Director, Microsoft Asia Pacific for their contributions to the discussion.

[2]R&D magazine (2011) 2012 Global R&D funding forecast. December. www.rdmag.com.

impact of a policy might rather be strongly determined by the way it is implemented.

With the exception of Greece, all European states increased their investment in R&D in 2012, including Italy, Ireland and Portugal who faced major economic burdens during this time. The 7th European research framework programme (FP7) and Competitiveness & Innovation Programme (CIP) covered the 2007–2013 period and aimed at doubling the investments to reach Euro 60 billion in fundamental research by 2014. The goal was partially achieved in some domains and disciplines. Horizon 2020, the 2014–2020 Research and Innovation programme is now under deployment. However, the US still leads in a number of advanced technologies.

MOTIVATIONS BEHIND THE DEPLOYMENT OF GLOBAL STRATEGIES

Both governments and companies, especially multinational corporations, advocate for increased efforts in developing innovation capabilities in the emerging countries. In a 2011 study, the consulting firm McKinsey looked into the behavior and motivations of multinationals regarding their investments in R&D in the emerging countries.[3] Two-thirds of the surveyed executives declared that their companies invested in R&D in those countries. In doing so, most of them tried to align their R&D policies with the global objectives set out by their group, whether these were to capitalize on the lower development costs in the emerging countries or to have better access to consumer insights. However, many top executives in the headquarters of these organizations seemed somewhat disappointed with the performance of their research facilities in the emerging countries. In particular, they referred to the problem of the high specialization in specific fields, which is not conducive to a transversal approach to innovation and provides limited opportunities to combine service value and technological value. Why then do they invest in R&D in emerging countries? For 44% of these managers, the primary motivation is to gain access to low-cost research and solutions even if this motivation factor diminishes in importance over time. For 39% of them, the main motivation is to have better access to consumer insight and to seize local market opportunities. By getting closer to their future customers, companies hope to learn to

[3]McKinsey Global Survey (2012) R&D strategies in emerging economies. Figures and field, March 2011.

operate in a manner better suited and adapted to the local market. By developing their products in Asia, multinationals therefore hope to capture market share destined to increase in the future. Even more importantly, companies tend to deploy innovation strategies that take full advantage of the possibilities provided by the permanently evolving global mapping of innovation and technology development.

Mapping the Global Innovation Landscape

North America and the European countries have switched from public to private investments in R&D. Increasingly; they rely on commercial firms in the global innovation race.[4] Direct government funding for research is in sharp decline since the 60s, leaving private firms to pick up the slack. The ratio of R&D spending on GDP decreased by 50% in the US; and from 0.31% in 1986 to 0.17% in 2009 in Britain — the lowest level in the G7 in contrast to the growth in Japan, Germany and South Korea. In the US, this public-to-private switch has been relatively successful, even if the overall ratio has fallen behind South Korea. The problem is with Europe, with a weak ratio at 1.9% (1.1% for the UK), despite Germany's efforts. On the other hand, a group of economists maintain that public benefits of radical innovation are likely to be higher than the private ones.[5] Therefore, genuinely significant new inventions end up diffusing, despite intellectual protection systems, and copycats and competitors eventually benefit. Three key issues significantly influence research spending: Local Tax treatment of R&D, merger policy, and pressure from shareholders to rein in R&D spending.

John Kao, author and consultant, pointed out in 2007 that United States has churned out innovations after innovations all over the 20th century. However, many countries, from Singapore to Finland, China to Chile, are placing innovation at the top of their national agenda since the beginning of the 21st century. Companies' executives can now weigh different national approaches to innovation in terms of their firms' strategic requirements. In doing so, they can practice innovation arbitrage taking advantage of differences in regulatory environments as well as in the cost

[4]The Economist (2012) America and Europe are relying on private firms in the global R&D race. Aug 25th.

[5]P. Aghion, J. Van Reenen, L. Zingales (2009) Innovation and institutional ownership. National Bureau of Economic Research. Issue 14769. And: N. Bloom, M. Schankerman, and J. Van Reenen (2013) Identifying Technology Spillovers and Product Market Rivalry. Econometrica, Vol. 81, No. 4. pp. 1347–1393.

of talent, specialized services, and the inputs to the innovation process.[6] Professor Arnoud de Meyer, President of the Singapore Management University, defines the six keys to ecosystem advantage: Pinpointing the added value; structuring different partner roles; stimulating complementary partner investments; reducing transaction costs; enabling flexibility and co-learning; and finally engineering value capture mechanisms.[7]

Facilitating the globalization of innovation management, a set of driving forces is suggested by John Kao. First, the rise of innovation as a currency of global competition has been observed, along with the emergence of the international style of management heavily influenced by values of openness, exploration, and risk taking, with English as a working language. To our understanding, however, situations in Japan and China might have differed in recent times. Second, a global war for talent is taking place, where scientists, designers, and entrepreneurs can be attracted from anywhere by appropriate incentives, professional opportunities, and creative ambiance. Third, many countries have placed innovation right at the top of their national agenda. Numerous nations now systematically develop innovation strategies, infrastructure, and human capital initiatives. Finally, fast growth of networks allow entrepreneurs to access resources on a worldwide basis, allowing for the emergence of a new type of brokers linking talents and assets separated by location or language.[8]

By setting up a lab or a marketing office, or partnering with strong local players, in a promising environment where talent is concentrated and resources are readily available, international companies can benefit in innovation hot spots by availing themselves of capital, R&D tax credits, regulatory relief, and special facilities. Firms can also position themselves as "systems integrators", picking and choosing from the best practices and resources across the globe and combine them in new ways. They can then practice "innovation arbitrage" by taking advantage of differences in regulatory environments as well as in the cost of talent, specialized services, and other inputs to the innovation process. Some forms of national innovation system may be identified.[9] A so-called 'focused-factory

[6] J Kao (2007) *Innovation Nation, How America Is Losing Its Innovation Edge.* New York: Free Press.

[7] A De Meyer (2012) Keynote speech. *Cross-Border Innovation & Entrepreneurship Symposium.* Singapore: Singapore Management University and ESSEC Business School, December 3rd–5th.

[8] J Kao (2009) Tapping the world's innovation hot spots. *Harvard Business Review,* March.

[9] *ibid.*

innovation model' combines clear strategic intent with concentration of infrastructure and high-level talent in an effort to discover and deploy new solutions to ambitious challenges. Singapore and Denmark, for instance, focus their innovation investments on a handful of industries and research fields.

Singapore is committed to funding scientific research, and increased its R&D funding in life sciences, clean technology, and digital media from 2.6% to 3% of GDP by 2010. Companies involved in R&D projects in Singapore can cover up to 40% of their expenses as part of a funding from the government. Biopolis was created in 2003, as a major biomedical research center. It has plans to employ 4,000 researchers by 2015, in proximity to 6,000 scientists in another R&D park Fusionopolis specializing in research in material science, clean technology, and digital media. Multinational companies such as GlaxoSmithKline, Novartis, Takeda, Valeant and CombinatoRx have committed to investments in Biopolis, where resources are shared under a unique relationship among government agencies, venture capital firms, academic research labs, and others. In wireless technology, precision manufacturing, or clean technology, Finland and Denmark would be the hot spots. Emerging economies such as Chile and Vietnam have invested in focused factories to attract manufacturing companies from other markets.

For Kao, the 'Brute Force Innovation Model' stands as a system where massive amounts of low-cost labor and capital are poured in a portfolio of innovation opportunities. China and India for instance, but also Brazil, hope that their enormous ideas will yield in valuable discoveries. Then, the 'Hollyworld model' provides opportunities to build a 'global creative class', attracting more and more entrepreneurs to gather in one place. Silicon Valley in the 1990s, and more recently Bangalore, Helsinki, or Toronto may be considered as adopters of such a model. India, on its side, is shifting its role as the world's service back office to that of innovation leader. The Indian Institutes of Technology have proved instrumental at establishing working connections with counterparts in the US and in Europe. Successful entrepreneurs of Indian origin, who have succeeded in the West, are cementing ties with their homeland in globalized technology enterprises. India is also managing the talent pools of other countries as part of its global strategy. For example, moving up the innovation value chain, Tata outsources some of its IT development assignments to Chile and Ecuador.

Finally, several countries have developed end-to-end innovation systems in support of an overall national agenda. These 'large scale' models combine stewardship mechanisms, funding bodies, research institutions, and structures for business and academic collaboration. Finland's innovation system was designed, in part, as a response to the catastrophical economic change in 1991 due to the collapse of the Soviet Union. Today, its Prime Minister chairs the Science and Technology Policy Council. Stimulus funds are allocated flexibly and across a range of public and private players, *via* Tekes — the national industrial R&D fund — and Sitra — the national innovation fund. Sitra subsidizes initiatives such as user-centered designs, new kinds of health care services, food and nutrition, or energy conservation. Aalto University, the "innovation university" started in 2009 with $1 billion in new funds, now focuses on technology for wellness and quality of life.[10]

By mobilizing a global network of resources, firms may prove worthy of optimizing the allocation of human resources and will develop more agile approaches toward development. As systems integrators, organizations might address a series of critical questions regarding the most attractive places to implement innovation efforts. As additional research units emerge and new technological advancement blossoms in various countries, the global innovation landscape continues to evolve. Emerging economies, especially in the Asia Pacific region, are demonstrating a growing interest in innovation policy, considering the development of disruptive innovations as strategic directions.

Benefiting To and from Emerging Economies

According to a report by the National Science Board published in 2012, North American multinationals are increasingly locating their research operations overseas.[11] All the innovative and high-performing companies conduct R&D operations in Asia. The move benefits Asian economies, which, the report insists, increase their own investments in science and technologies. The number of researchers employed by US multinationals overseas has doubled between 2004 and 2009, to reach a total of 270,000. China and India are the regional destinations preferred by multinationals for outsourcing their R&D.

The rationale for relocating R&D operations to Asia and investing in research there has shifted the focus from costs to competencies: Multinationals seem to no longer be looking for lower costs, as was the case in the past, but rather skills, competencies and talented individuals.

---

[10] *ibid.*
[11] www-networkworld.com (2012) January 23rd.

Certain multinationals also declare they invest in R&D in Asia to obtain disruptive and/or frugal innovations that, increasingly, emerge out of Asia.[12] Companies therefore increase their presence in Asia to capture market opportunities; thanks to disruptive innovations that could radically modify the markets in years to come.

In the last 50 years, Asian economies have grown astonishingly quickly. However, in doing so, they only produced a limited amount of innovations of their own. Today, the situation seems to have reversed. More and more significant innovative solutions originate in Asia, whether they are the product of multinationals working in Asia, or of Asian companies themselves. R&D in Asia is evolving: It is now just a matter of adapting existing products for local consumption than implementing true product development processes.

Organizational Zoom...

STMicroelectronics R&D in Asia; Thales research center in Singapore

STMicroelectronics, the premier European player of semiconductor packaging, has based its R&D capabilities dedicated to the assembly of semiconductors, what the company calls packaging, within close proximity to its clients and business units on the one hand, i.e., in Europe, and on the other hand, near the manufacturing ecosystem mainly in Asia.

Thales Group took this evolution into account. Their initial model, which mainly consisted of offering their products in Asia in a way that was adapted to local markets, soon revealed the limitations. Thales then decided to establish a proper research center in Singapore, thereby benefiting from local technologies, competencies, and laboratories, in a structure that was fully integrated within the group's research organization and its technology. That is how products were developed in Singapore and are now offered throughout the world, beyond the Asian market, in Mexico, for example.

Going from success to success, the Singapore research team is receiving more and more support from the group. First established in 2003, this team now has 60 full-time employees, scientists, and engineers, half of which are involved in the CINTRA program, launched in partnership with the CNRS and the Nanyang Technological University. Their research is dedicated to signal processing and sensors, high-accuracy

[12] The Economist (2012) Frugal ideas are spreading from East to West. March 24th.

systems simulation environments, as well as some target markets in the areas of defense, smart cities (transport, security), and air traffic management.

As mentioned by The Economist Intelligence Unit in 2012, moving R&D to Asian locations also helps in capturing market opportunities. By 2016, it is expected that Asia (ex-Japan) will contribute to 33.8% of global GDP, *versus* 13.2% in 1990 (measured using purchasing power parity).[13] Another significant element in the R&D strategies implemented by multinationals is that they perceive there are centers of excellence in Asia which can address the needs of emerging markets in general. To succeed in Latin America or benefit from the announced growth in Africa, for example, firms will need specific solutions that are frequently being developed in Asia. It has become clear to companies that the innovation processes deployed in emerging markets are different from the processes in mature markets; such processes in the emerging markets do not seem to be adapted to the economy of tomorrow. By conducting R&D in Asia, multinationals are therefore forced to re-design their innovation processes in their own countries, whether in the US or Europe.

If in reality, R&D centers in Asia are still more dedicated to product development than basic research, this situation has to change. In fact, the most advanced countries, such as Japan, Singapore or South Korea, have started investing significant amounts into fundamental research. To conclude this brief overview, we probably should underline two observations. The first is that, increasingly, in survey after survey, the criterion of adequate skills and competencies among Asian engineers and scientists has become the decision-making factor to motivate the multinationals in locating their R&D units to Asia. For this reason, certain regions that were dominant in the past in terms of innovation, such as Bangalore or Beijing, are no longer seen as relevant implantation locations. It has become too difficult and too expensive to recruit there. New regions are emerging in Asia, where multinational companies have better access to pools of highly qualified and talented researchers. The other factor is that fiscal incentives and other advantages offered by Asian governments for R&D investments are less attractive to foreign companies than policies implemented by these countries to support higher education and to promote the acceptance of new technologies by society.

[13]The Economist Intelligence Unit Limited (2012) Coming of age: Asia's evolving R&D landscape. Commissioned by Mercer.

Developing Advanced Technologies as Nations' Priority

Governments in Asia, as well as leading regional companies have realized that product improvement and incremental innovation, even if key for productivity increase, are not sufficient to prosper in the 21st century in the context of global competition. The ability to shape disruptive innovations is critical for the survival of firms. Disruptive technologies, i.e., the technologies that make previous ones obsolete, are a crucial lever for growth. According to Harvard Professor Clayton Christensen, who has introduced the concept of Disruptive Innovation, companies that do not invest in breakthrough, rule-changing solutions are doomed to disappear within a short period of time. François Luc of Thales Group has said that "investing in technology is the life-insurance for the future of your company!" As opposed to incremental innovation, disruptive innovations are characterized by much sustained development and by the fast achievement of high level of performances. However in their initial phase, disruptive innovation strategies are usually less attractive since their user friendliness quotient has to be worked out.

One of the difficulties of running disruptive innovation projects, compared with incremental developments, lies in the complexity of assessing their ROI. If this assessment proves actually doable, it might mean that it is no longer about disruption. New business models have already been installed by competitors, who took the spot on the market before you do. Moreover, it proves very difficult for a team working by itself to produce a disruptive innovation, because technologies that are not directly connected to the organization's core business have to be mobilized. In developing new generations of mobile phones, for example, organizations have to work with experts from battery technologies. A company can seldom bridge the technological gap, that will allow it introduce a high-performance disruptive innovation, alone.

How then should a technological innovation be managed in order to ensure its success? An important concept to take into account at this point is the level of technological readiness. There are many scales available to measure this. The one used by Thales has nine levels, and is divided into two major activities: Research and Technology (R&T) on the one hand, and engineering on the other, i.e., product development. In the first field, R&T, teams work on principles (principles of Physics, Algorithms, etc.) without any link to a particular market. If you are not active in this first field, you will only be able to offer incremental innovations, and, as discussed previously, your future will be limited to 30 years. In the second field, the focus is on refining products to increase their performance, by applying the principles defined by the other group. This goes through different phases of prototyping and introducing the new technology to the market and users, and the risk is progressively reduced. It is therefore always crucial to identify in which phase of the process you are, to guide the project accordingly.

For an articulate company, it is usually feasible to find institutional support and financing for sound basic research. Once the innovation is marketed, you have access to customers and you are expected to achieve decent profit margins. The difficulty arises during the intermediary stage: The testing phase. This is the phase when the company may have cash-flow difficulties, and is often referred to as the "valley of death". The challenge lies in being able to transfer the innovations from fundamental research toward the engineering phase, by walking through the "valley of death" unscathed, and to transform your innovation into a marketable product. Is it better to go through the "valley of death" alone or with others? In other words, should you adopt open or closed innovation?

Open innovation helps to limit the risks inherent to disruptive innovation. In order to push ideas through the innovation funnel, from basic research to engineering, you should be capable to feed it externally, through mergers and acquisitions, with spin in and spin off, alliances, and research collaborations, etc. The open innovation model progresses faster than the closed model. Some tests will be successful, while others will fail, but at least the company would have shared the investments and their risks. In return, however, this means accepting a more reduced market share. Partnerships can also be detrimental as they can present a threat in terms of intellectual property or can generate knowledge leaks. Open innovation, however, remains an excellent way to maximize the chances and to minimize the risks of an innovation process. Some Asian countries provide a favorable ecosystem for successful innovations.

TOWARD OPEN AND DISRUPTIVE INNOVATION IN ASIA

It is important to remain humble in terms of innovation. As François Luc at Thales said: "You sow without ever being sure of how much you are going to reap. To a certain extent, it all depends on the soil you chose, it is therefore important to identify the right territory." Open innovation requires several ingredients: Multidisciplinary universities; growing research budgets; quality of life; multicultural openness; appropriate protection of intellectual property, and finally a proximity with end users and the market that make it possible to implement changes and obtain faster development rates. A certain number of Asian countries, Singapore in particular, offer these qualities.

Grooming Talents in Science and Engineering

As discussed previously, investments made by multinationals toward R&D in Asia were initially motivated by a competitive cost environment. Today, this is no longer their main attraction lever. Within the context of global and open innovation, what seems much more attractive for these multinationals, and now one of their main criteria for locating R&D, is the possibility of close proximity to advanced competencies in Asia. For example, it was in Asia that STMicroelectronics further enhanced its technical expertise to face up to one of its major technological challenges: Switching from gold cables to copper wires in the design of its semiconductor packages. With the ever-increasing price of gold, it became paramount for the company to implement this change promptly, as it was a condition for its competitiveness. The group has been working on this transformation for the past several years, with a significant acceleration in recent times. By end 2012, it had already converted 800,000 km of wires used in its packages to copper, i.e., to the moon and back! To implement this change, it worked with its material suppliers (wires, resins, etc.), the wafer manufacturers to redesign the chips, its equipment suppliers (located mainly in Asia), and with qualified teams (engineers, technical experts, etc.). However, it was more difficult to find additional technical experts in packaging in this field in Europe while there are several of them in Asia; thanks to the presence of large, well-implanted manufacturing networks.

Will the younger generation continue to be attracted to research and develop their competencies? Or will they be drawn to the Sirens of finance? Xavier Baraton, from STMicroelectronics, defends that in order to nurture this breeding ground for talent; it is essential to foster closer ties with universities, to identify and attract future researchers early on and show them the extent of the value of the R&D developed by the company. Future researchers take interest in high-tech projects, even more so when they realize that significant investments are allocated to these projects. It motivates them. Another key motivational factor for young scientists and engineers in this industry is that they have the opportunity to discover a wide variety of different dimensions which the profession has to offer, and that too in a variety of countries. It may therefore also be an opportunity for them to discover the world.

In this respect, the pedagogic content and approach of scientific and engineering schools are evolving, and the focus now is to develop an education system that fosters a sense of innovation. There may well be less researchers and engineers tomorrow, but there will be more people who contribute toward innovation mechanisms. And given the crises that banks have experienced following the financial crisis, we may expect the strong financial 'magnetism' to actually wane a little bit. Microsoft receives more and more applications for research, mainly in Asia. The number of European candidates who express interest in working for Asian research centers is increasing. This will ensure a higher diversity of competencies in these centers. Statistics corroborate these points of view. The US provided less than 4% of the young engineers who graduated in 2011 worldwide. In North America and the UK, people may still be encouraged to pursue careers in better-paid sectors, such as law, finance, and medicine. An important fact to bear in mind is that 65% of scientific PhD students in the US are no longer American nationals. At the other end of the spectrum, 34% of the newly minted engineers in 2011 graduated from Chinese institutions. Asia has somewhat taken on the responsibility of producing a higher number of engineers and scientists than western countries. France and Germany are more successful than the UK in maintaining a certain appeal for scientific studies and engineering tracks. But Asia remains the most fertile location, and is also the place where engineering schools today have the most significant resources.

Proximity to Next Generation Users and Key Value Chain Links

Asia is undergoing a considerable mutation in the field of technological innovations. Schematically, we can consider that in the past, the countries with a mature economy launched technologies that others then adopted and possibly adapted. Today, however, innovative and diversified uses are emerging throughout Asia. It is crucial for any innovative company to follow these mutations as closely as possible, in both usage and behavior. That is the concern expressed by Microsoft. The near absence of Microsoft research centers in Asia in the past led it to miss market and/or technological opportunities, such as the localization of mobile phones for the Chinese market or certain touch technologies.

A presence in Asia is all the more necessary for companies like STMicroelectronics, as the electronic components sector is mainly based there. The group needs to maintain close relationships with its materials and equipment suppliers who are based mainly in Singapore, Malaysia, Japan, and China and in the Philippines. In 2011, 70% of electronic components were made in Asia. As for the customers for whom the components were destined to be integrated in mobile electronics for example, 63% of them were based in Asia, 24% in Europe and 14% in the US. However, 33% of them were actually for Asian companies themselves, 38% were European and 29% were North American. In other words, large US mobile phone manufacturers, for example, produce in Asia, but design their innovations in California. The same can be said about components in general. For that reason, it is important that companies such as STMicroelectronics are present where the products are designed, i.e., in the west.

Obstacles to Overcome

Despite these benefits, some obstacles inherent to the Asian context must be addressed for open innovation to be successful. For example, although Asia offers many competencies, it is still more difficult for companies to retain engineers. It is a challenge for high-tech multinational companies to continue to reinforce their attractiveness and retain the qualified professionals. The expertise of Asia-based R&D units, such as STMicroelectronics, even though advanced, could be further strengthened. Innovation in Asia is still in its emerging phase in the field of microelectronics. Asia has been home to development and manufacturing processes for 40 years, however, it is only in the last 10 years that products have been designed in the region. In the field of microelectronics and certainly in other fields as well, the investment framework in Asia is somewhat less structured and coordinated than in Europe. While in Europe there are communities of investors, in Asia things are done on a country-by-country basis. In this scenario, Singapore has proved to be a highly attractive place, where its Economic Development Board offers important funding and has established a fruitful partnership with companies such as STMicroelectronics, and helped to launch its unit in Singapore.

Another question that regularly comes back to the fore is that of intellectual property, especially in China. However, it is believed that this issue will be less and less the subject of debate in the future. How do multinationals see it? From the point of view of Thales, for instance, which only has a small team in China; this is hardly a problem as its research activities here have proved to be winners. This fast-paced development makes it possible to hold an upper hand — whenever you are faster than others and enter markets before others, the difficulty in protecting intellectual property disappears. STMicroelectronics has just established a packaging R&D team in China, in addition to its manufacturing infrastructure in Shenzhen area with experienced and trusted Chinese team members who have worked with the group for a long time. Above all, patent protection might still prove the best way to safeguard intellectual property rights. For Microsoft, in addition to implementing an efficient legal department, it split its projects between different countries, to limit the risks in terms of intellectual property. The true challenge for the software industry resides less in the protection of intellectual property than in fighting against the marketing of counterfeit products.

A lot remains to be done in this respect, both from a legal point of view and in terms of education. Partnerships are possible solutions for the future, but in any case, it is a long-term endeavor. The positioning of Chinese companies regarding intellectual property has evolved drastically over the last 20 years. 20 or 30 years ago, China had few patents of its own. This trend has now been reversed. It is in the interest of the system as a whole to protect intellectual property. Some companies like IBM decided to invest in collaboration with researchers from top Chinese universities from early on to co-develop technologies and co-own patents.

Organizational Zoom...

Microsoft R&D in Beijing

Beijing is home to Microsoft's largest R&D center in Asia, with 230 researchers, among the 900 that the group has in its 13 research centers. Every year, the group invests nine billion USD in R&D, of which 500 million are destined to the center in Beijing. It is specialized in natural user interfaces, next generation multimedia, data intensive calculations, on-line advertising and computer science fundamentals. In addition, Microsoft has 27 centers in the Asia–Pacific region, where small teams develop R&D activities, on a project-by-project basis.

Every 10 years, Microsoft launches an innovation to reinvent itself. This rhythm has fundamental consequences on how research, engineering, sales, and marketing have to work together. And the stakes are important! One company out of five of the Fortune 500 ranking uses Microsoft's cloud productivity services; Microsoft's SharePoint collaboration platform is used by more than 70% of the companies from that list; its Xbox Live has no less than 40 million users in 35 countries around the world, and Skype has reached 200 million users every month. Since Windows 8 was released, more than 40 million units have been delivered, more than the total of Mac users. Today, thanks to the evolutions of Windows, Microsoft is reshaping its company around a new vision.

Microsoft is primarily known as a software company. However, its survival depends on it going beyond this domain. While software is what made the difference in the 1990s, the integration of software and hardware is becoming increasingly important. Microsoft must therefore evolve toward hardware, and a large number of hardware manufacturers are located in Asia. If the company is not one step ahead on how it intends to implement its vision in hardware, it could miss a major technological shift. A new era is dawning. An additional challenge to comprehend is that software is becoming increasingly complex, because they have to integrate cloud computing and be used on a very wide range of devices. Finally, services should also be taken into consideration: We have to offer new applications, and make sure they are fully integrated.

Microsoft celebrated the 10th anniversary of its Beijing research center in 2011. The company supports the work of top Chinese academics and encourages its researchers to publish their work.

AMBITIONS, POLICIES AND PRACTICALITIES FOR INNOVATION IN CHINA

The oldest known Chinese classic texts, usually called the *I Ching* or book of changes, lays down a system of symbols assisting scholars in their attempt to identify order in what otherwise seems like a set of chaotic events. It is said that the principles originated with the mythical legendary ruler of China Fu Xi (2852–2738 BC). The text positions creativity as indeed a higher-order quality, which might be the preserve of a few who are 'awakened': "Creativity comes from awakening and directing men's higher natures, which originates in the primal depths of the universe and are appointed by Heaven (...). The Creative knows the great beginnings. The Receptive completes the finished things".[14] As epitomized in a 2008

[14]R Wilhelm (translator), CF Baynes (translator) and H Wilhelm (preface) (1950) *The I Ching or Book of Changes*. The Bollinger Foundation, New York, NY 1967: Princeton University Press, Princeton.

book on the transformation of society and business in China, there is no shortage of creativity in the Middle kingdom, as opposed to a frequently mentioned myth.[15] As Philip Kotler states: 'The Chinese people are very entrepreneurial'. Didn't Deng Xiaoping's modernization plan take its root in individual initiative and entrepreneurial drive?

Back to the Future: A Tradition of Creativity and Innovation

Two key achievements mark the early stages of the achievements of the Chinese people: The emerging of writing around 4,600 BC, and the discovery and production on a massive scale of bronze for weapons and ritual vessels. Around 3,000 BC, the people who settled in the Yellow River valley were already farming, and using pottery, wheels, and silk. By the 4th century BC, the Chinese had developed a lodestone compass to indicate direction; then paper was invented in around 200 BC and the technology for deep drilling boreholes took off one century later. In AD 132, the Han dynasty scientist Zhang Heng is credited with the invention of the seismograph which was able to detect an earthquake at a distance of more than 800 km. Gunpowder appeared during the Tang period and firearms were already spread out during the 11th century and from the beginning of the 12th century, cannons were produced with barrels made from bamboo with iron rings.[16] The Song dynasty (960–1279) is a befitting example of a creative technology-driven period with the invention of movable-type printing, and water clocks, among other advancements.

The China maritime power is less remembered since it mainly characterized the search for world trade domination by the Ming emperor Zhu Di in the beginning of the 15th century. To signal his power to the world, the Chinese emperor commissioned the construction of 250 leviathan nine-mast large boats; they were called 'treasure ships' because of their ability to carry vast quantities of goods. Records say he also ordered the building of 1,350 patrol ships, 400 warships, and another 400 cargo ships that would transport water, grain, and horses for the fleet. In 1405, Admiral Zheng He was sent on multiple overseas trips with almost 30,000 crews, soldiers and merchants on a fleet of 200 vessels aiming at dazzling and intimidating

[15] Kunal Sinha (2008). *China's Creative Imperative, How Creativity is Transforming Society and Business in China*. Singapore: Wiley.
[16] *ibid.*

foreign rulers. The key objective was to draw them into the tribute-to-China system in exchange for trading privileges and protection, establishing a vasality system on the basis of trade and influence rather than brute force. By 1433, the "Grand Eunuch" Zheng He had been on seven voyages, visited more than 30 Asian and African countries, and brought back animals and goods, later depicted in graphic art, for emperor Zhu Di.[17] As a commercial strategy, as opposed to a military one, for world domination, it was unheard of at the time and unmatched in its domination and execution.

In the late 18th century, China began to dominate world trade, but the introduction of the "Western powers" in the region in the 19th century, coupled with the weakening of the old imperial order contributed to a dramatic reshuffle of the cards. The first part of the 20th century stands as a period of suffering and conflicts for China. The People's Republic experimented with varied economic policies in the 50s and 60s with limited success in terms of technological development. The cultural revolution of the 70s proved disastrous at many levels and resulted in the death of millions due to starvation and "rectification" measures.[18] After Chairman Mao Tse-tung died in 1976, the newly reshaped Chinese Communist Party implemented a revitalizing economic policy under the guidance of Deng Xiaoping. By the end of the century, China had recovered part of its past manufacturing power and was ready to challenge American and Japanese leaderships in technology development.

China's Attempt to Lead in Knowledge Creation and Technology

For years, except for a limited number of large state-owned companies, the greatest proportion of surplus generated by the sales of products manufactured in China primarily benefited foreign companies. For instance, the analysis of the value-added associated with the commercialization of Apple products unveiled the fact that manufacturing items had provided little to Chinese suppliers in comparison to the profits made in the US. Manufacturing has generated enormous trade surplus but the government also realized that the Chinese Yuan's inevitable appreciation would make

---

[17]G Menzies (2002) *1421 — The Year China Discovered the World.* London: Bantam Books.

[18]JK Fairbank (1987) *The Great Chinese revolution 1800–1985.* New York: Harper Perennial.

China's low-tech exports uncompetitive. Eventually, low-and-middle manufacturing processes would have to shift to countries such as Indonesia, Malaysia, Thailand, and Vietnam. In 2006, Beijing decided to dramatically increase the number of created-in-China technologies. Equally important is the fact mentioned by former US Secretary of State Henry Kissinger that there is some sort of belief, on the Chinese side, that their country has suffered a century of humiliation, "and that it's now their turn to demonstrate at least equality".[19] However, based on major statements, the Chinese government appears to fully recognize that long-term cooperation with the West is vital.

Since the announcement by President Hu Jintao in 2007 that "Innovation is the core of our national development strategy",[20] China has quietly moved toward a new stage of development, shifting its focus to advanced technology development. The country has become the world's second largest producer of scientific knowledge. As explained by Thomas M. Hout and Pankay Ghemawat, cajoling, co-opting, and often-coercing Western and Japanese businesses have proved a key component of the country's strategy.[21] The government also engaged in a plan to increase R&D expenditure from 2010 level of 1.7% of GDP to 2.5% by 2020. Finally it has funded mega technology projects a little similar to what the French government did during the late 60s and 70s with the nuclear power program, fast trains, and aerospace technology plans. Foreign organizations are encouraged to bring their ideas to China but that has to directly deliver value to the country, and their operations are circumscribed by government's goals. Multinational companies operating in strategic sectors, such as air transportation, power generation, high speed rail, information technology, and possibly electric car, have to share their technologies with state-owned Chinese enterprises and research laboratories as a condition to compete for a share of the world's fastest-growing market.

The Chinese economy expanded by 9% a year from 2008–2010 but slightly slowed down from 2014. In 2010, it surpassed Japan to become the second-largest economy in the world. The Government expresses its

[19]Bloomberg BusinessWeek (2011) Charlie Rose talks to Henry Kissinger. January 24th.

[20]GM Hancock and G Krompacky (2007) The rise of China: Changing patterns of global innovation and entrepreneurship. *Walter H. Shorenstein Asia–Pacific Research Center*, January 1st. http://aparc.stanford.edu/news/the_rise_of_china_changing_patterns_of_global_innovation_and_entrepreneurship_20070101/.

[21]TM Hout and P Ghemawat (2010) China *vs.* the world, whose technology is it? *Harvard Business Review*, December.

willingness to significantly strengthen innovation, particularly in energy, transportation, the environment, agriculture, information, and health. The efforts aim at boosting the production of proprietary intellectual property and at applying advanced technologies to public services and urban management. Finally, the country also looks at modernizing its defense capabilities, especially with its aerospace and maritime programs. Accordingly, Beijing plans to introduce unique technical standards that would reduce dependence on imported technologies by 30%, and ensure that China will rapidly become one of the world's top five economies according to the number of patents granted and scientific papers published. Several mechanisms to achieve these goals can be identified that rely on: Tax incentives; increased investment in strategic areas; reshaping of procurement policies to favor local technologies; incentives for multinational firms to transfer their knowledge. By creating a tipping point at which foreign companies have to locate their most-sophisticated R&D projects and facilities in China, the government expects to eventually catch up with the US as the world's most technology-driven economy.

In parallel with government efforts, the entrepreneurs of China have proved extremely active over the past 10 years, tapping on their knowledge of the domestic market, their capacity for quick adaptation to market changes, and their resourcefulness. In a rich study of entrepreneurship and innovation in China, Yinglan Tan explains how 46 Chinese firms have succeeded to have their IPOs listed on NASDAQ from 1999 to 2011. Based on a series of interviews, the book describes the opportunity that Neil Shen, co-founder of Ctrip Capital China, seized upon for a Chinese travel site; how Ray Zhang, leader of eHi, scaled up one of the most advanced hybrid car-rental companies in China; how Wu Jianguang, founder of Really Sports, created one of China's largest retail outlet of sports products by dramatically enriching the customer experience. Other stories include the start of Semiconductor manufacturing international Corporation by Richard Chang in 2000; the transformation of Neusoft Corporation by Liu Jiren into the largest IT solutions and services provider in China; the launch of a leading video sharing site Tudou.com in April 2005 by Gary Wang; or the introduction of the venture capital industry in China by Tan Lip-Bu, founder and chairman of Walden International, among others.[22] Most of China's firms are focused on process innovation while

[22]Yinglan Tan (2011). *Chinnovation, How Chinese Innovators are Changing the World.* Singapore: John Wiley & Sons (Asia).

still using existing technologies in product creation. Typically, Chinese startups first adopt business models from foreign origin before shifting to implement their own ideas.[23] While turning threats into opportunity, innovating Chinese entrepreneurs look both inside the country and outside for disruptive innovation and successfully implement collective model of entrepreneurship.

In the beginning of 2010, much of the debate was whether China was more likely to become innovative or would remain a fast follower of foreign leading companies.[24] Indeed, at that time, China was suffering from a research culture that focused more on quantity over quality and the use of local assessment standards. The country appeared to be lagging behind in becoming a true global innovation power.[25] However, companies in China were already significantly engaged in innovation with numerous promising developments. More recently, McKinsey Gordon Orr and Erik Roth, long time observers of the Chinese economy, have expressed their belief that the pace of innovation was quickening and a new spirit of innovation was spreading across sectors, including in government.[26] As John Oyler, CEO of the Chinese biotech company BeiGene, pointed out in 2013; 'anything is possible, we can make it happen, there is no challenge we cannot conquer, we will surprise the world'. For instance, in 2014, Beijing Genomics Institute (BGI) became the world's largest genetic-sequencing firms, and now claims to account for roughly 50% of global capacity, while developing some of the world's most advanced biologic-computing models. In 2013, Huawei has unveiled very innovative new smartphone platforms built on chips and software designed in-house; while Xiaomi, another smartphone producer created in 2010, is often compared to apple for its innovative marketing strategy and has sold more than 15 million pieces in 2013. Lenovo, the Chinese computer maker with 2014's first quarter revenue at $9.8 billion, very successfully addresses the global high-priced-laptop market, with its Yoga ultrabook controlling more than 40% market share

[23]K Hille (2011) Chinese entrepreneurs at a copycat stage. *Financial Times*, October 22nd.
[24]G Orr and E Roth (2012) A CEO's guide to innovation in China. *McKinsey Quarterly*, No. 1.
[25]AK Gupta and Wang H (2011) China as an innovation center? Not so fast. *The Wall Street Journal*, July 28th.
[26]G Orr and E Roth (2013) China's innovation engine picks up speed. *McKinsey Quarterly*, No. 3.

in the $900-and-above price band in the US retail market in 2013.[27] Also, the Lenovo table PC Horizon evolved from a one-person machine to a multiple-person machine — a family entertainment center focusing not only on hardware but also on applications.[28] The firm has implemented a smart international acquisition strategy, working hard to integrate the management of each newly acquired organization such as IBM server and Motorola phones. In several respects, says William J. Holstein, Lenovo may already have leapfrogged some Western multinationals and its success offers a definitive answer to the question whether Chinese companies can compete globally. They clearly can.[29]

Organizational Zoom...

Applied Materials; AstraZeneca; BASF; General Motors; Merck; PepsiCo; and Pfizer

In 2010, Applied Materials, a global leader in semiconductor-making equipment, transferred many R&D activities to China and relocated its chief technology officer there. In 2007, AstraZeneca established its Innovation Center in China as part of a $100 million investment made in the country with a special focus on cancer prevalence in Asia, such as gastric and liver cancers. The journey has just started and the company has recognized the significant qualities of both types of scientists, trained overseas or locally, even if in some very specialized areas, they are still in short supply.[30] BASF had 10 R&D facilities in Shanghai alone, before opening its new Euro 55 million pan-Asian innovation center and greater China headquarters in the city, where special incentives are available from the local government. The company expects to locate one-quarter of R&D staff in Asia by 2020. Since 2010, General Motors and its Chinese joint-venture partners sell more cars in China than in the United States. The car manufacturer's Advanced Technical Center in Shanghai, where team-based innovation efforts are encouraged, comprises several engineering and design labs.[31]

---

[27]McKinsey Quarterly (2013) Thriving in a "PC-plus" world: An interview with Lenovo CEO Yang Yuanqing, no. 3.

[28]McKinsey Quarterly (2013) Thriving in a "PC-plus" world: An interview with Lenovo CEO Yang Yuanqing, No. 3.

[29]WJ Holstein (2014) Lenovo Goes Global, China's most recognizable brand has plans to overtake Apple and Samsung. *Strategy + Business* (76), August 8th. www.strategy-business.com/article/00274?gko=abf3e-LenovoGoesGlobal

[30]S Yang (2012) Pharmaceutical innovation: AstraZeneca's experience in China. *McKinsey Quarterly*, No. 1.

[31]K Wale (2012) Automotive innovation in China: The view from General Motors. *McKinsey Quarterly*, No. 1.

Merck's Asia R&D headquarters was scheduled to become operational in 2014. In November 2012, PepsiCo opened its largest research and development center outside the US in Shanghai. The $45 million new facility includes kitchens where Pepsi chefs develop new flavors from traditional Chinese cuisine, laboratories where they taste-test them on consumers, and plants where prototypes are produced.[32] Pfizer moved its Asia headquarters also to Shanghai in 2010.

Understanding the State Strategy on Innovation

As discussed in detail by Thomas M. Hout and Pankaj Ghemawat in a 2010 paper,[33] the Chinese government introduced export subsidies for local businesses and has necessitated government agencies and state-owned companies to procure goods from Chinese-owned providers, if feasible. Also, a three-pronged plan has been implemented to contain foreign companies and enable Chinese firms to accelerate their production of advanced technologies. First, the State has ensured that it will be both buyer and seller in certain key industries. Second, it has consolidated several manufacturers into a few national conglomerates, expecting to benefit from economies of scale and concentrated learning. Third, the country requested multinationals to form joint ventures with key national players, allowing for the transfer of advanced technology in exchange for possible business opportunities. China was working on acquiring the most advanced technologies as fast as possible in order to benefit from the local and global infrastructure building booms under way, as well as to act while the national currency was still undervalued.

In an attempt to identify which pattern may prove successful, several experts have tried, over the years, to compare research policies as implemented in the United States with what China does. For instance, R&D funding and R&D execution are significantly differently managed, with the US heavily relying on Industry for the funding whereas China depends on government investments, for instance. Paradoxically, the result of a more recent study shows that more than 80% of Chinese believe the US is staying ahead of China on innovation while only 40% of Americans seem to share that opinion.

---

[32]P Waldmeir (2012) China offers a taste of R&D to come. *Financial Times*, November 13th.

[33]TM Hout and P Ghemawat (2010). *Op. Cit.*

Table 9.1 Comparing R&D Funding and Execution in the United States and China

R&D Funding (2009)		R&D Execution (2009)	
United States	China	United States	China
Government 29%	Government 69%	Government 10%	Government 23%
Industry 65%	Industry 21%	Industry 71%	Industry 67%
Academic/NP 6%	Academic/NP 10%	Academic/NP 19%	Academic/NP 10%

Sources: TM Hout and P. Ghemawat (2010); China *vs.* the world, whose technolgy is it?; Harvard Business Review, December. Batelle Institute (2009): Global R&D funding forecast, December 22nd; China Economic Quarterly (2006), Q3; and OECD R&D statistics (2009).

The State sometimes also coordinates its efforts to stimulate growth in a particular industry by implementing new regulations on foreign players in the related sector. Hout and Ghemawat quote several examples from the energy sector. From 1996 to 2005, international wind energy equipment makers held 75% share of the given Chinese market. The government made the decision to dramatically grow that market by offering buyers large new subsidies and other incentives. At the same time, it quietly increased the local-content requirement on wind turbines from 40% to 70% and substantially hiked the tariffs on imported components. As the market exploded, foreign players were unable to meet the increased demand. On the other hand, the Chinese manufacturers, usually licensing technology from small European turbine producers, expanded rapidly and cost-effectively. By 2009, Chinese companies led by Sinovel and Goldwind, ended up controlling more than two-thirds of the market while foreigners did not win a single central government-funded wind energy project from 2005–2010. In the solar panels equipment, prices fell dramatically in 2009 and 2010 mainly due to hyper-competition between Chinese producers, which spilled into international markets. As a consequence, Germany's pioneering company Q-Cells, but higher cost player, slid from an operating profit of 16% of sales in 2008 to an operating loss of 60% in 2009. In 2010, Chinese companies, such as Suntech, Yingli, and JA Solar, using new techniques to improve the efficiency of solar panels,[34] exported 95% of their production and controlled a third of the US market and half of the German market.

Another insightful illustration can be found with the capital-intensive sector of the high-speed railway systems. At the beginning of the 2000s, the technological superiority of corporations, such as Alstom, Kawasaki, and Siemens, allowed them to control about two-thirds of the market. These multinational players were subcontracting the production of components to state-owned companies while delivering end-to-end systems to China's railways operators. In 2009, the government began looking for foreign organizations willing to bid on railway projects to form joint ventures with the national equipment producers CSR and CNR, with a maximum 49% equity stake in the new companies for the foreigners. Then, multinationals were asked to share their latest designs, and 70% of each system had to be made in China. Only a few sophisticated components, accounting for 15–20% of the value, are imported. The combination of large experiment, low manufacturing costs and advanced technologies created national champions, such as CSR and CNR, who are well equipped for the $120 billion international market, including developing countries where China funds modernization projects, and also developed markets, such as Australia and New Zealand. At the end of 2014, the announcement of the merger between CSR and CNR may precisely reflect Beijing's objective to form mega-sized state-backed conglomerates to address international markets better. Named CRRC Corp. with total sales of up to $32.25 billion and more than $13 billion net profits in 2013, the new firm is expected to tap into opportunities in Southeast Asia, Central Asia and Eastern Europe.[35] In 2014, the two companies controlled almost 100% domestic market share in railway locomotives, bullet trains, passenger trains, and metro vehicle but only commanded 2.2% of the international market.[36]

Organizational Zoom...

Huawei; Haier; JAC; and Sany

The telecom powerhouse Huawei, a private firm, created or acquired 16 R&D centers around the world. The firm claims that 44% of

[34] G Orr and E Roth (2012). *Op. Cit.*
[35] wantchinatimes.com (2015) Merger of China CNR, CSR announced. January 3rd.
[36] J Yang (2015) Train giant's merger signals SOE shake-up. *South China Morning Post*, January 1st.

> its 140,000 employees in 2012 were in R&D, many of them in its research campus in Shenzhen.[37] Its founder, Ren Zhengfei, served as an engineer in the People's Liberation Army like many of his fellow entrepreneurs in China. For auto manufacturers, Turin is the place to be, with JAC, FAW, and Chang's operating R&D centers there. Machinery manufacturer Sany, whose main international competitors include Caterpillar and Komatsu, initially attempted to succeed in the European and US markets by relying on homegrown talent and technology. But a few missteps encouraged the firm to establish R&D centers closely tied to its European and US headquarters and to staff them with professionals from those countries. Sany's 2012 acquisition of Putzmeister, Germany's leading cement pump maker, gave the company access to a onetime competitor's technology.[38] Haier, a leading appliance and consumer electronics manufacturer founded in 1984, totaled $29.5 billion in revenue for 2013. Its CEO, Zhang Ruimin, a former Red Guard during the Cultural Revolution, is credited for having turned around a little-known, bankrupt refrigerator manufacturer into one of the world's largest multinational controlling 10.2% of the global market for large home alliances in 2014. Mr. Zhang now claims to be ready to eliminate the firm's entire middle management.[39] The company also develops a wide network of global designs and R&D facilities in the US, Japan, Korea, Italy, the Netherlands, and Germany. In the age of digital transformation, Haier home appliances have become the symbols of smart family strategy,[40] providing smart and connected appliances such as air Rublik Cube, "clean-free" washing machine, Haier Boguan fridge, and Tianzun air conditioner.

Naturally, foreign companies and the Chinese government have been disputing the new set of measures and some firms have reassessed their Chinese strategy. On the one hand, multinational companies work at reducing security risks in relation with their technologies. On the other hand, many firms try to identify development strategies which they intend to run in China to strengthen their positions in the global marketplace. Also, Chinese R&D units no longer have the advantage of low cost. National players may face some challenge in their efforts to catch up. For instance, Aviation Industry Co. of China (AVIC) that launched a promising commercial airliner in 2014, the wind energy companies Sinovel and Goldwind which benefited from their outstanding domestic growth,

---

[37]The Economist (2012) Who's afraid of Huawei. August 4th.
[38]TM Hout and P Ghemawat (2010). *Op. Cit.*
[39]The Economist (2013) Chinese industry: Haier and higher. October 12th.
[40]heraldonline.com (2014) 6th consecutive world top — Haier large home appliance brand retail takes 10.2% of global market in 2014. December 31st.

and the rail-transport equipment CSR and CNR even if effectively merged, still had to prove their ability to compete with Western, Japanese, and South Korean leaders in terms of technological advancement. So, despite reservations regarding the Chinese government policy, foreign R&D centers multiply in the country; fewer than 30 foreign units were noted in 1999, against more than 600 R&D centers by 2004. By 2010, that number had more than doubled, and their scale and strategic importance had increased.[41] For the last five years, a number of initiatives have been taken in that domain despite the slowing down of the economy.

The nurturing of talents also attracts increasing attention in China with huge public and private resources allocated to the development of world-class universities and most advanced academic research programs. The 1,300-year-old imperial examination system, abolished in 1905, has incentivized the establishment of a new school system in China.[42] The 1920s saw the growth of a number of national universities, such as the prestigious Peking University, in addition to the setting up of a dozen of Christian colleges, like Fudan University in Shanghai by the Catholic missionaries, among others. In addition, there were a few private Chinese institutions funded from wealthy sources, such as Nankai University in Tianjin as the personal achievement of Chang Po-ling, an alumnus from Columbia University, New York. The Rockefeller foundation took the lead in contributing to China's higher education. Tsinghua University, a product of American indemnity payment, Peking University, Jia Tong University system especially in Shanghai, National Central University, and, finally at the apogee of research, the Academia Sinica emerged. A creative set of private institutions also emerged, such as Yenching University, incorporated in New York, St. Johns University, and Peking Union Medical College, among others.[43] They were all Sovietized in the 1950s and were almost entirely destroyed during the Cultural Revolution.

The system has been rebuilt with energy and determination along the past 30 years. The transformation of China's postsecondary education system proves more dramatic than anything else seen before in North America and Europe. However, even if the country has more than 2,400

[41] M Regina, M Abrami, WC Kirby and W McFarlan (2014) Why China can't innovate, and what it's doing about it. *Harvard Business Review*, March.
[42] B Keenan (1977) *The Dewey experiment in China: Educational Reform and Political Power in the Early Republic*. Cambridge: Harvard University Press.
[43] JK Fairbank (1987). *Op. Cit.*

universities and research centers, not all of them are getting upgraded equally. In order to satisfy to the strategic goal of turning China into an innovation-driven country by 2020, the government has chosen 10 of its leading universities to receive enormous funding to achieve world-class status. To lure international senior researchers to national institutions, the government has launched a series of schemes since the mid-1990s, followed in 2011 by the "foreign-expert thousand-talent scheme".[44] Leading Chinese universities are now recruiting world's top faculty in priority fields. The number of students graduating from Chinese universities has grown from 1 million in 1999 to 6.7 million in 2012. That year, 23.9 million students attended institutions of higher education in China, four million more than in the US, with private schools accounting for more than a quarter of all institutions. Nature, which is the globally recognized leading scientific journal, has published 303 papers by Chinese scientists in 2012, up from 46 in 2006. Protection of intellectual-property has being given real attention by the Chinese Justice department.[45] The authorities have announced the handling of 2,347 cases in 2012, up almost 40% from 2011. Finally, as emphasized by many observers, more Chinese graduates move into the commercial sectors than into government; and it is anticipated that fast-growing firms will more likely breed innovative practices than the large-scale, mostly state-owned, companies of the 1980s or 1990s.

PERSPECTIVES

Mutations in information technology and connectivity that the world is experiencing have reached such proportions that they announce the beginning of a new era for the economy and pose challenges for innovation. These mutations, particularly significant in the Asia–Pacific region, are partly due to the proliferation of devices, combined with the desire of consumers to live an ever-better experience. A personal computer is a crucial collaboration tool for the actors of the information society, but individuals influenced by their experiences in the private sphere, aspire to more advanced and interconnected tools and applications in their professional activities: Television, tablets, and especially smart mobile phones. This proliferation is reinforced by cloud computing, making it possible

[44]The Economist (2014) China: Higher education, a matter of honours. November 22nd.
[45]Wang Xiaodong (2013) Progress in fight against fake products. *Chinadaily.com.ch*, January 23rd.

for individuals to access their information from all sorts of devices, even from remote Asian locations. And while the border between private and professional spheres is getting increasingly porous, especially in countries such as South Korea or China, organizations must create their own social networks. Finally, companies have more and more remote workers, which force them to address new challenges in terms of security of information and for the management of application portfolios.

Impact of Technology Mutation

More generally, the technology mutation has also made consumers more social; they no longer contact a brand when interested in a product, but rather join in conversations about it on virtual networks. They have very precise opinions on the technology that a company might offer them before they even talk to its representatives. Studies unveil that 44% of the consumers who complain, do it directly *via* social network sites, and no longer complain directly to the company. Companies must learn to listen to these new communication modes and most importantly, should act on them. Another important change is the big data explosion, which is multiplied by 10 every five years. This massive amount of data comes from more and more diverse sources: Images, ads, publications on social networks, information from mobile phones, etc. This transforms the technological approach companies may take.

For Microsoft, for instance, these mutations imply four major challenges. First, the company has to touch and integrate technologies that are still in their infancy. Very soon, screens will be embedded in credit cards. Flexible screens offer truly novel capabilities in terms of user experience. The second element is about the importance of cloud computing, which also in its early stages. By 2016, the number of data centers throughout the world would have doubled. This poses fundamental challenges in the management of energy, security and storage, and the understanding of what a data center is. The third direction for innovation has to do with social companies. The question is now raised about the experience being provided: Is it only about creating more data, or is it about creating data that people can act on? The fourth challenge is about big data. For example, how to manage massive amounts of data that open doors in terms of marketing and customer relation management?

Health has been identified as one of the main market drivers for innovation, even for actors who *a priori* seem to bear no relation to this domain. That is the case of many Internet companies, which consider that their solutions should contribute in different ways to improving the lives of individuals, even if it creates vital challenges. Microsoft, again, engaged in the fight against tuberculosis, a critical disease that can be healed with a six-month treatment. Nine million people are infected every year, and 1.4 million die because they did not follow the treatment to the end. The company has introduced a simple technology in India, which can be implemented in a large number of cities, and provides the capacity to monitor nine million people. This solution is based on the connection between a biometric system, specific software and a cloud computing service. Every time a patient comes for his/her medication in a medical center, he/she is identified. Those who do not show up are contacted. Within the first six months of operations, it is estimated that 500,000 people have been saved; thanks to this service.

For STMicroelectronics, health has also been identified as one of the main vectors of growth in the coming years, all the more so, considering that consumers are ready to pay for applications in this field. There is less pressure in terms of costs for these types of applications compared to traditional consumer goods. The company already offers chips capable of performing DNA analysis, or monitoring insulin pumps. In a related field, STMicroelectronics has also vastly invested in dependency. The Micro Electro Mechanical Sensors (MEMS), for example, transform movements in electric signals and can, for example, detect when an elderly person falls at home. This type of solution is bound to experience significant growth. With an aging population in major Asian countries such as China, Japan, and South Korea, a younger more well-being conscious generation in India, the Philippines or Vietnam, health represents huge potential market opportunities for IT providers and other firms. The challenge is then to successfully combine the different dimensions of high-technology medical equipment with the possibility for doctors to intervene remotely with the flow of medical data and the protection of confidentiality. All these elements are important fields of innovation for the future.

It now seems unavoidable not to integrate sustainable development into innovation policies. This challenge is particularly pressing for the energetic consumption of data centers, for example. It takes approximately 37 million liters of water for the cooling of a data center that turns on 20 MWH, and about 10 million liters of this water cannot be recycled. For countries already afflicted with water shortages in Asia, it is a true loss. One of the innovation areas is to reduce its water consumption. Microsoft, for instance, is working on a new generation of zero carbon data centers. They will be placed near waste treatment facilities, so they can recycle the biomass and use wind and hydraulic power. In April 2012, Microsoft inaugurated its very first center of this type. A couple of years ago, the Power Usage Effectiveness (PUE) index was created. In the last 10 years, Microsoft's data centers have moved up in value, from 2 PUE to 1 PUE, considered the best value in the world. The group was rewarded for its results in sustainable development for the third year in a row in the US by the agency that delivers the Electronic Product Environmental Assessment Tool (EPEAT) eco-label. This offers new innovation perspectives. If Microsoft and the recipients of the EPEAT award decide to establish more data centers in Asia, which is likely, they will have to do so while also respecting the environment.

Health, sustainable development, and mobility are varied domains and are equally important. They force us to take a more global and open approach to innovation, by mobilizing a wide range of actors in an ecosystem-based approach: Engineers, economists, psychologists, and sociologists, etc. Each and every one of these actors has a role to play in combating these societal challenges. They will have to learn to talk the same language, despite their different competencies and cultures. Moreover, many innovation-related challenges will not be solved by a mere technical approach. Innovation is not pure science. To be able to imagine new concepts or technologies is a definite quality, but to be able to transform these concepts into value-added products that will find their place on the market requires higher level competencies that are seldom taught in scientific and engineering schools. One of the largest challenges for universities is precisely to combine competencies that vary in nature. This is the condition for the development of innovations that are truly adapted by the societies of tomorrow.

In the future, we will therefore witness the emergence of an integrated approach to innovation. Multinationals will establish innovation hubs, in which marketing, engineering, and other business units will be able to exchange ideas with one another as well as with other external actors, so as to better understand the needs of the market. Countries like Finland or Israel have developed many such hubs. MIT's Mobility Lab acts as an innovation hub in the industry of mobility, by bringing together professionals from the automotive industry, urban planners, etc. These new research entities must be able to address societal issues and benefit from the value derived from the market, research, and higher education.

Let a Million Enterprises Bloom in Asia

There is no statistical evidence that suggests a significant relationship between firms' financial performance and the level of their spending in R&D, especially in service environments, on the contrary innovation leaders, as perceived by markets, demonstrate strong results.[46] For companies, factors such as their capability to successfully run innovation projects, create breakthrough new products or processes, and reshape their business models, are more significant than the amount of cash poured into their R&D pipeline. This means carefully articulating their talent architecture and incentive policies as well as efficiently mobilizing the massively interconnected network of organizations, technology developers, and consumers' creativity pool that make up the contemporary economy. Equally important is the ability of firms to take advantage of the "innovation models" deployed worldwide and then to practice what has been introduced as "innovation arbitrage" in order to maximize their chances of success. Historically, as we have seen, a number of radical innovations have originated from Asia, especially from China. For the last 50 years however, the region has been credited with only a few bold innovations. But the Asian landscape is rapidly reshaping with a significant provision of technological advancement and new value equations.

Today, it is assumed that economic growth and development, measured in terms of sustainable and shared increase in per capita income, are driven by entrepreneurship and the deployment of innovation. As a consequence, governments all over Asia, as well as international development agencies worldwide, extend recommendations, incentives, and regulations to boost entrepreneurship and stimulate innovation. However, Japan, Singapore and, surprisingly enough Hong Kong, arguably the most developed countries with the highest GDP per capita, unveil the lowest level of entrepreneurial activity in the region, while the Philippines with more than 39% of TEA (total early stage entrepreneurial activity)[47] is dominating the landscape, followed by Thailand with almost 20% and Indonesia with 19%. Despite pro-business government policies and well-educated work forces, entrepreneurship spirit is lacking especially in Singapore and Japan due to people's

---

[46] B Jaruzelski, V Staack and B Goehle (2014) The 2014 global innovator 1000: Proven paths to innovation success. *Strategy&* (PWC). http://www.strategyand.pwc.com/global/home/what-we-think/global-innovation-1000?utm_source=VanityURL&utm_content=innovation1000&utm_medium=redirect&utm_campaign=inov1000.
[47] *The Global Entrepreneurship Monitor.*

continued attraction to careers in public service, and the cultural and social norms which do not support risk taking and creativity.

Confucianism emphasizes frugality, respect to the authority and hard work, but traditionally places little value to the merchant class and the business community. On the other hand, Buddhism stands for a higher level of tolerance toward failures, which proves in favor of any entrepreneurship journey as displayed by the impressive new-businesses-driven economic advancement in Thailand for instance. Malaysia, an Islamic country, enjoys the advantages of a vibrant entrepreneurial spirit with a growing importance to women in small and medium sized enterprises (SMEs). However, the country still displays a low perception on entrepreneurship as a career. Even if the role of SMEs in Indonesia proves very significant, innovation-driven entrepreneurship may be limited by high taxation and burdensome government requirements.[48] The most developed countries usually had more entrepreneurial activities in the past. For instance, South Korea experienced astonishing growth during the 60s and the 70s, mainly driven by favorable tax incentives supporting the expansion of family-controlled SMEs which have today become large multinational giants — Samsung, Hyundai and Posco. Japan, which shaped its manufacturing industry with market-friendly policies over the last 60 years, now lacks young and risk-taking entrepreneurs partially due to demographic and cultural issues.[49]

As mentioned, European multinational companies and the US are rapidly increasing their innovation related investments in Asian countries, while local governments are emphasizing the importance of higher education and research for sustainable development. R&D spending continues to grow while globalization accelerates, driven by Asian economies, especially China together with South Korea, Taiwan, and Singapore. Corporate Japan, still a leader in technology advancement, has however slowed down its R&D investments in recent times. While the key motivation to conduct R&D in Asia used to focus on access to low cost resources, tax breaks, and frugal innovation formulas, the importance of customer insights is now proving a key priority for MNCs. Accessing competencies and well trained talents also stands as a key motivating factor, but organizational challenges are proving

[48]A Maritz, V Le, S Masli, A de Waal, B Verhoeven and Shieh Chich Jen (2013) Entrepreneurship and growth in emerging mini-dragon economies. A South-East Asia perspective. *Chinese Business Review* 12(1), January. pp. 41–47.
[49]H Mathe (2013) Asia's cultural challenge. *The Business Times* (Singapore), Nov 15th.

very real. China and India remain the most favored regional destinations for R&D efforts, but multinationals suffer from staff turnover problem and the relative inconsistencies of performances compared to Europe or North America. Despite this, the role of Asia in global R&D will only increase since combined public and commercial efforts to stimulate innovation are unequalled in the rest of the world. Also, more centers of excellence for new product development will emerge in the region to capture market opportunities with benefits for other regions from this learning.

Asia-based firms are simultaneously increasing their R&D capacity at home as well as in western countries aiming to equip themselves with "global innovation brains". China, on its way to becoming the world's largest economy, legalized entrepreneurship in 1979 and created the Special Economic Zones (SEZs) where commercial firms have flourished. Despite the Confucian cultural heritage, entrepreneurship is now thriving at a very large scale encouraged by public policy and the necessity to maintain growth at an annual rate of 8% or above. The most successful country in reducing poverty rates, China now focuses on turning itself into a knowledge driven economy. "Innovation is the core of our national development strategy" said President Hu Jintoa as early as in 2007 — "homegrown innovation is (...) the path to sustainable economic prosperity and global leadership". From 2006–2011, China's public spending on research and development increased by more than 20% a year, reaching US $164 billion by 2012 with more than 217,000 new patents issued.

However, when it comes to introducing rule-breaking solutions and products, Chinese firms are lagging, while economic growth is significantly decelerating. Even if the *I Ching*, previously introduced, positions creativity as a "higher-order quality", for many observers, including Lee Kai-Fu from Innovation Work, "The Chinese education system makes people hard-working, teaches people strong fundamentals, and makes them very good at rote learning" (...) "It doesn't make them creative original thinkers".[50] It might be time for China, and possibly for other Asian communities as well, to question if the "infrastructure investment-driven" innovation should be complemented by a "creativity and experience-driven" form of innovation. In 2015, state-owned funds will supposedly control more than US $12 trillion in asset, a huge figure beyond the means of any one or group of private investor(s). Commercial firms are also flexing their muscles. Following their

[50]L. Hilgers (2013) Lee Kai-Fu, China's Innovation Idol. *Bloomberg Business Week*. August 22.

own models, Tencent and Weibo, for instance in the online service sector, are growing at an unprecedented speed compared to anywhere else in the world; while Huawei and ZTE are ranked among the world's top five corporate patent registrants in recent years.

In India, the leading 40 public research centers, extensively collaborating with industry, have produced "more US patents than all domestic Indian private firms combined". And despite the not-so-convincing Tata Nano experiment, reverse or frugal innovation originated in Asia which dramatically alters the way technological development is addressed in the west. "Radical cost-cutting" now impacts the design of almost any product, from medical devices, to tractors (following Mahindra) or wine cooler (following Haier). In Singapore, where Prime Minister Lee Hsien Loong is also Chairman of the Research, Innovation and Enterprise Council, R&D efforts still originate from the government that also raises required investments, despite the economic slowdown. Public resources are successfully used to incentivize firms to move into industries considered as strategic, such as biotech, clean energies, or media-entertainment. Significant funding also allows public universities to pursue ambitious science programs while wooing foreign talents. Naturally, the community of scientists, engineers and designers, in public and private sectors, has continued to grow dramatically over the years. However, numerous observers questioned the ability of the private sector to sufficiently stimulate customer and market-driven innovation, since government policies, even when well refined, can only provide limited support.

As manufacturing matures in South Korea, services are upgraded and nurtured, the government introduced measures to provide a boost to a series of strategic service sectors and instill a sense of innovation in services.[51] In the mid-1980s, the largest service sector in South Korea used to be an inefficient retail trade, now surpassed by a very active knowledge intensive business service industry. The information technology sector has firmly positioned itself as a growth engine for the Korean economy along the 2000s, with its GDP portion increasing dramatically. The rate of penetration of telecommunication devices and the development of the Internet service industry is dramatically high, along with the multiplication of e-commerce

[51] H Mathe (2013) Keynote speech: "Service innovation for economic growth, job creation and the improvement of quality of life". *2013 International Forum on Service Sector Advancement*. Seoul, Republic of Korea: Korea Development Institute and Minister of Strategy and Finance. Services as a new engine of growth, November 7th.

platforms. Also, Korea's franchising sector has rapidly developed over the last 15 years; its growth being backed by public authorities in view of propelling leaders overseas. Five key service activities emerge for Korea, which are of strategic importance: Health care, Education, Green finance, Content and software, and finally Tourism and events organization. South Korean authorities insist that there is a greater development potential of the service sector relative to manufacturing in terms of employment creation, market expansion, and export opportunities.

Chapter 10

INNOVATING TOWARD SUSTAINABILITY*

Alvin Toffler once said that we cannot predict the future. Similarly, we can borrow a dictum from the illustrious co-founder of Southwest Airlines, Herb Kelleher, which says that "the only thing that we can truly plan for is change: An axiom that goes hand in hand with the necessity of innovation". Change is the foundation on which company survival is built and businesses today have to be agile, reactive and creative in order to remold and rebuild their enterprises in response to and anticipation of an unstable environment, to develop new product and service offerings and to foster transformative progression in a volatile and fiercely competitive globalized environment. A number of academic studies have concluded that volatility at the firm level increased somewhere between two and four-folds from the 1970s to the 1990s.[1] In today's markets, nothing stands still for long; product life cycles are perpetually decreasing today and even the most successful firms may be supplanted by better, faster, savvier entrants to tomorrow's markets. Indeed, Schumpeterian theory rings true: The assertion that competitive advantage will become increasingly more difficult to sustain.[2] Today, the difficulty in maintaining competitive advantage has become so convoluted

*The author would like to thank Ziying Huang for her significant contribution to this chapter.

[1]DD Sull (2009) Competing through organizational agility. *McKinsey Quarterly*, December.

[2]RR Wiggins and TW Ruefli (2005) Schumpeter's ghost: Is hypercompetition making the best of times shorter? *Strategic Management Journal* 26(10), pp. 887–911.

that it has fallen into the realm of "hypercompetition"[3] and this is a phenomenon that is seen across a broad range of industries.[4]

In addition, our society's awareness of our collective impact on the planet has been growing over the last several decades. Numerous reports, studies, articles, theses, and books have been published documenting impacts and opportunities, for example, dangers of certain chemicals,[5] rapidly expanding human population,[6] resource constraints,[7] anthropogenic climate change,[8] and the business opportunities for those who are aware of the sustainability issues.[9] A most frequently cited work is the report '*Our Common Future*' by the Brundtland Commission published in 1987. In the report, the commission coined what has become the most often-quoted definition of sustainable development: "Development that meets the needs of the present without compromising the ability of future generations to meet their own needs".[10] The report emphasized that sustainability is a three-legged stool of people, planet, and profit. Organizations should balance all three dimensions to ensure that all processes, products and manufacturing activities adequately address current environmental concerns while maintaining a profit.

Indeed, companies today are facing "hypercompetition" as well as the pressing challenges of sustainable development. To achieve sustainable growth, the only way is through innovation. Innovation has been unequivocally established as an imperative for business success in the 21st century, and is of key importance in the quest for the elusive and sustained competitive advantage. Innovation also enables the firm to prepare for the

[3] RA D'Aveni (1994) *Hypercompetition: Managing the Dynamics of Strategic Maneuvering*. New York: Free Press.
[4] RR Wiggins and TW Ruefli (2005). *Op. Cit.*
[5] R Carson (1962) *Silent Spring*. Boston: Houghton Mifflin Company.
[6] DH Meadows, EI Goldsmith and P Meadow (1972) *The Limits to Growth, Vol. 381*. London: Earth Island Limited.
[7] RB Gordon, M Bertram and TE Graedel (2006) Metal stocks and sustainability. *Proceedings of the National Academy of Sciences of the United States of America* 103(5), pp. 1209–1214.
[8] S Solomon *et al.* (2007) "IPCC, 2007: Climate change 2007: The physical science basis". *Contribution of Working Group I to the Fourth Assessment Report of the Intergovernmental Panel on Climate Change*. Cambridge: Cambridge University Press.
[9] B Willard (2012) *The New Sustainability Advantage: Seven Business Case Benefits of a Triple Bottom Line*. Canada: New Society Publishers.
[10] Brundtland Commission (1987) Our common future, report of the world commission on environment and development. Published as annex to General Assembly document A/42/427, United Nations: Development and International Co-operation: Environment.

future, however unstable and unforeseen that may be. Yet, while most managers acknowledge the necessity of innovation, and the change that it entails, putting innovation into practice is fraught with challenges.

SUSTAINABLE VALUE CREATION THROUGH SERVICE INNOVATION

Most companies now acknowledge the critical importance of innovation for sustainable profitability, and yet, the concept of innovation is nonetheless fuzzy, unpredictable, and perceived as risky by many companies. Innovation is undoubtedly a key priority everywhere, with a huge corpus of knowledge, theories, and experiments, but results are often disappointing. Indeed, the multi-stage process of innovation, i.e., from idea generation to the profitable implementation of ideas is decidedly complex. The "secret sauce" of innovation is the product of organized processes and intangible ingredients, which is something that many companies have successfully mastered, while others have much to learn.

For businesses to become more competitive and gain significant advantages in the current global environment, they must innovate their business models and astutely engage with their customers to create compelling value.

"The Right Business Model" to Sustain Competitive Advantages

In a marketplace full of commodities, companies everyday face new difficulties in maintaining competitive advantage and differentiating their offer. In the sea of imitators, fast-movers and innovators, sitting pretty on a cash cow is rarely an option. Technological innovations have a lifespan and, more often than not, time will eventually render a once revolutionary concept obsolete. Incremental innovation can only take a product so far; at some point in time, its very existence will be thrown into question. As a case in point, the Sony Walkman, once a revolutionary product, was successfully upgraded to the Sony Discman when CDs came to replace the traditional tape. When the digital revolution arrived, effectively changing the way music was transferred and sold, Sony was too reliant on its classic formats to adapt to the digital format which was quickly superseded by Apple's iPod. The book world is currently facing the same challenge, i.e., the electronic book format is challenging the very business model of the traditionally bound book.

Companies that innovate their business models are more likely to achieve a sustainable competitive advantage, as it is much more difficult for a competitor to copy a business model than it is to just replicate products. According to Alexander Osterwalder and Yves Pigneur, authors of *Business Model Generation*,[11] a business model describes the value an organization offers to various customers and portrays the capabilities and partners required for creating, marketing, and delivering this value and relationship capital with the goal of generating profitable and sustainable revenue streams. Business model innovation involves the design and management of innovation at the organizational and systems levels of business. To come up with the "right business model", companies must rethink the business approach, and focus on the unique offering, servicing clients, re-designing the business and creating value. In addition, innovation requires constant thinking, planning, and experimentation, to create new capabilities and to successfully implement a new business model.

Innovation that enables the company to change, adapt and to reinvent its business model is imperative for company's survival and growth. It is no longer sufficient to compete on the back of a better product. Companies must do a better job at innovating their business models that integrate product innovation with process and service innovation. IBM's global CEO study conducted in 2006 indicated that firms innovating their business models exceeded the growth of firms that were innovating their products, services, markets or operations.[12] Results of IBM's survey are outlined in Figure 10.1.

Today, business model innovation is perceived as a way to increase profits but also a way to avert threats from competitors. The iPod developed by Apple outshone other MP3 players that were basically the same. One key factor of Apple's success is that it developed a complex hybrid product–service offering that elevated user experience to a new level, which has taken competitors a long time to come, even near to copying. Other notable examples include: Dell's production to order system, Zara's supply chain for fast fashion and the Toyota Production System. In each case, companies redesign the elements of their business, and refine their business models

[11]A Osterwalder and Y Pigneur (2010) *Business Model Generation*. New York: Wiley, August.

[12]IBM Global Services (2006) Business model innovation — the new route to competitive advantage. http://www-935.ibm.com/services/uk/cio/flexible/enflex_wp_business_model_innovation.pdf.

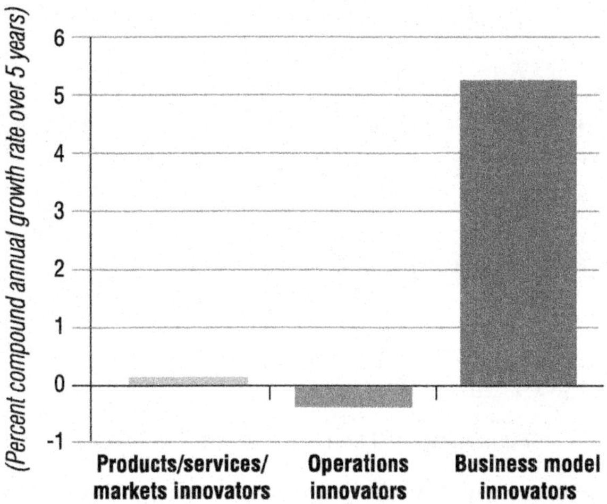

Figure 10.1: Operating margin growth in excess of competitive peers (Source: IBM Global Business Services. Expanding the Innovation Horizon. The Global CEO Study 2006.)

to create competitive advantage, differentiation in the marketplace and deliver enhanced value to customers. It is particularly difficult to imitate improvements that add value for customers. Companies should seek and consider improving their business models at all times.

Organizational Zoom...

Dell, the Early Years: Creating Value through Business Model Innovation

"This belief — that by working directly with customers we could get them technology faster, provide a better level of service, and provide better value — was the basis of the business. The fundamental business system was quite powerful and delivered lots of value to our customers. We screwed up lots of things, but the one thing we got right was this core business model, and it masked any other mistakes..."[13]

— Michael Dell, founder of Dell

The computer hardware industry become a highly competitive industry by mid-1990s where a number of consolidations through mergers and

[13]M Dell, C. Fredman (1999) Direct from Dell: Strategies that Revolutionized an Industry. New York: Harper Collins Publishers Inc.

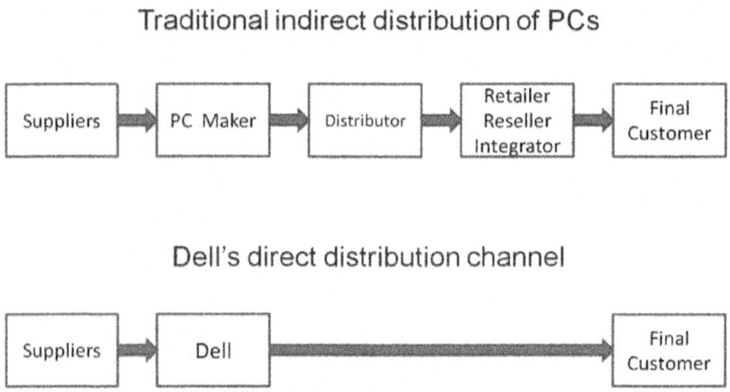

Figure 10.2: The Dell Distribution Model.

acquisitions have occurred, but also divestment of existing businesses, such as IBM making the strategic decision to discontinue manufacturing and marketing of desktop and laptop computers. Dell, however, was willing to rethink its business model and develop a new approach to remain competitive in a highly dynamic business environment. Dell's initiative is of historical value in today's situation, and perfectly illustrates what business model innovation might mean for a company, and sometimes for an entire industry.

Dell implemented a customer-driven, build-to-order model that replaced the traditional build-to-stock model of selling computer through retail stores (see Figure 10.2). Specifically, the company focuses on the following elements of the business:

- Direct customer relationships.
- Direct sales.
- Customer segmentation for sales and service.
- Build-to-order production.

The Dell business model was refined by removing those elements and activities that did not create value for Dell or its customers. The above figure compares the Dell business model with the traditional personal computer distribution model.[14]

The business model innovation implemented by Dell has provided the following advantages for the company:

- Increased market share for personal computers.
- Ability to deliver a "made-to-order" customized PC quickly to the customer.

[14] J Kapeleris (2013) Business model innovation. An industry catalyst white paper. QMI Solutions. Eight Mile Plains, Australia. March.

- Increased client satisfaction.
- Reduced cost through minimizing inventory.
- Increased profit by utilizing a direct sales strategy.

Dell did not bring any improvements to the technology of the personal computer, but it did combine both suppliers' and its own organizational/ distribution system innovations to deliver compelling value to the end users. One of the critical elements of Dell's success was the way it organized its value chain. Dell's competitors were incumbents who had difficulty in replicating Dell's strategy, as selling direct to customers would upset their existing channel partners and resellers. Dell, as a new entrant, had no such constraints.

Another important factor of Dell's success was the choice of products it sold through its distribution system. Over time, Dell developed dynamic capabilities around deciding which products to build beside desktop and laptop computers, and has since added printers, digital projectors and computer-related electronics. Of course, the whole strategy depended on the availability of numerous non-captive suppliers capable of producing at very competitive prices.[15]

Customer Experience: A Link to Sustainable Value

Today, most innovating companies have changed their business models. Instead of focusing on geographies, local markets and products, they dynamically focus on customer experiences and problem-solving. If focusing on competitors leads strategists inexorably to the notion of sustainable competitive advantage, focusing on the customer leads them to the notion of value.

Innovation today has unquestionably entered the waters of "open innovation". Companies not only have a greater degree of connection with their consumers and their ecosystem, there is also a greater degree of submersion into what we call "useful universes", the environments in which we find the users dwelling and where tests can be conducted. The underpinning experimentation within the "useful universes" goes hand in hand with the services-oriented logic of today's business environment where consumer need is increasingly complex and sophisticated and integrated service-product solutions are defining the path to competitive advantage. The hazy boundaries between product development hovering in a tug-of-war between the consumer and producer will revolutionize the traditional

[15]DJ Teece (2010). Business models, business strategy and innovation. *Long Range Planning* 43(2), pp. 172–194.

push–pull dynamics of the market by irrevocably shifting the power toward a predominantly consumer *pull* tendency. Business problems today tend to get expressed and engineered in terms of satisfying the consumer: Products and services may thus be defined as "*Market's calling*".

Hence, understanding the ephemerality of market needs, and innovating to keep pace with them, is imperative for today's market. Research suggests that more than 86% of new product ideas never make it to the market.[16] And of the products that make it to the market, there is a failure rate of between 50 and 70%.[17] Research shows consistency in the cause of these high failure rates which are due to the uncertainty around, and lack of understanding of customer needs and values, and the fact that new products that failed in the marketplace did not meet any or enough unmet customer needs.

However, truly understanding the customers is extremely complicated. In the context of the fast-paced and protean marketplace, companies can no longer blindly pursue an "ivory tower" approach to new product or service development with their senses closed to the needs of the consumer. Innovative companies are astutely engaging with the consumer through "living laboratories" and dialoguing with the user. Orange has embraced the "usage laboratory" approach with its experimentation lab, enabling this dynamic company to acutely understand the importance of tests in semi-real environments. Transportation, banks, and other sectors are embarking on similar trends. The very notion of going out into the world and testing in real situations is the new *vademecum* of innovation. In the services economy, prototypes often have to be made that can connect with real "living" situations. The R&D laboratory that develops and tests products in isolation has reached its saturation limits in an economy of services.

The bottom line is the importance of putting into operation an approach that gets as close as possible to the final beneficiaries and is

---

[16]Based on a survey of 13,000 product launches from 700 industrial companies in the years from 1976 to 1981. See Booz Allen & Hamilton (1982) *New Products Management for the 1980s*. New York: Booz Allen & Hamilton. A survey by the Product Development and Management Association in 1995 suggested that to generate one successful product it took 6.6 ideas. See A Griffin (1997) PDMA research on new product development practices: Updating trends and benchmarking best practices. *Journal of Product Innovation Management* 14, pp. 429–458. See also RG Cooper (2001) *Winning at New Products: Accelerating the Process from Idea to Launch.* Cambridge, MA: Basic Books.

[17]See R Tucker (2002) *Driving Growth through Innovation: How Leading Firms are Transforming their Futures.* CA: Berrett-Koehler Publishing.

called "Living" innovation. Therein lies the new gold mine, where lots of new things shall be learned, fueling the development of market-oriented offerings that are acutely in tune with the consumer and designed to delight! It is arguably a more sophisticated evaluation of latent customer needs than traditional market research is capable of producing. With a deeper understanding of the consumer and the possibility to bounce ideas, interactive experience labs can be rich in nuanced discovery. Indeed, the "usage laboratory" is a propitious place for exploration, given that, by definition, it changes all the time!

Human-centered innovation is also a vector of social change and innovation gives it the means to construct a more dignified human life of shared prosperity. The Grameen Bank is a classic example of micro-credit that circumvented the norms of conventional banking by only lending to those who have assets or financial collateral; the Grameen Bank favors the poorest in society who do not have the means to reconstruct their lives. Other examples of social innovations include the brand of basic products, such as water, eggs, etc., under the brand of "ONE" of which profits are donated to charities in their entirety.

In addition, the stakes are mounting for successful innovation with figures suggesting that the products representing some 70% of current company sales will be obsolete in less than three years[18]; companies cannot afford to stand still and reap the benefits of current successes. As Patrick Pichette, CFO at Google, notes "*look at how much change has occurred in the digital space in just the last 48 months — 48 months ago, who knew about Netflix or Facebook*?"[19] The short lifespan of new products has become a determinant of their success, so mastering the processes for recurrent innovation is essential for companies today.

Hence, in this changing customer-relationship dynamics, there is a crucial need to manage the cycle of a new product and its service delivery over time. Change and adaptation are integral to satisfying the evolving customer need and the continuity of the relationship. The sense of timing that is critical to managing the adaptation of the offering over time is manifested in the reality of the shortening product life in the market. Today, the eventual obsolescence of products and services has become encompassed in their design, i.e., products and services are being permanently upgraded

[18]Deloitte (2005) Mastering innovation. Exploiting ideas for profitable growth.
[19]Manyika (2011) Google's CFO on growth, capital structure, and leadership. *McKinsey Quarterly*, August 2011.

and their design encompasses an acceptance of evanescence. The mobile phone device market banks on the perpetual and increasingly frequent upgrades by repurchase.

Apple broke new boundaries by creating a trend of repurchasing and multiple purchasing from its line of expensive, "must have" products such as the iPod, the iPad, and the iPhone. IKEA initiated this idea in the American market with the "Embrace Change" campaign. American consumers, at the time, viewed furniture as a once or twice in a lifetime purchase that predicated an insufficient market opportunity for IKEA. Thus, by encouraging consumers to embrace change and redecorate their homes, IKEA rewrote the needs of the entire market and created space for inexpensively priced Swedish furniture.

Organizational Zoom...

Medtronic: Evolving from Product Supplier to Solution Provider

Medtronic is headquartered in Minnesota. Its chairman and CEO, Omar Ishrak, has been instrumental in transforming the company from a medical device manufacturer to a premier medical technology solutions provider.

As the global medical technology leader in the management of chronic medical conditions such as cardiovascular diseases, pain, and diabetes, Medtronic is extending its cutting-edge technology portfolio to include focused, complementary services and solutions that treat patients beyond those who solely rely on or need implanted medical devices.

The idea is to closely align business objectives to the needs of hospital customers. This led to the company's launch of Hospital Solutions business in 2013. And in August, Medtronic announced that it had acquired NGC Medical, an Italian company that manages cardiovascular suites, operating rooms and intensive care units, for \$350 million.[20]

Earlier this year, L.E.K. Consulting published results of a hospital survey that showed that the second urgent need felt by hospitals, behind cost containment, is to improve efficiency. According to L.E.K experts, in Europe, the need for efficiency is even more acute in hospitals that have already travelled further into cost containment than their American counterparts.[21]

---

[20] A. Parmar (2014). Medtronics Finds Growth in Fledgling Hospital Solutions Business. http://www.mddionline.com/article/medtronic-find-success-hospital-solutions-business-November 18.

[21] *ibid.*

Medtronic's European division studied how it could improve the efficiency of technology delivered at the point of care, and it established the new business unit "Hospital Solutions". Soon the unit launched the catheterization laboratory ("cath lab") management program in which Medtronic and the hospital systems build in risk sharing for efficiency savings — guaranteed and underwritten — and clinical outcomes. The contract for the partnership outlines benefits and cost savings for both parties and hospitals, *via* the system of pay per procedure.

Specifically, in cath labs, many cardiac interventions are carried out using Medtronic products, including stents, which open up the clogged heart arteries with wire-mesh tubes. As part of the cath lab management program, Medtronic sets up a new cath lab within a hospital, coupled with the latest technology and equipment, and maintains the facility. The medical device manufacturer also deploys a Medtronic team to help the hospital implement lean six sigma programs to improve efficiency. Hospitals can choose from a menu of other Medtronic offerings, including[22]:

- Growth programs such as enhancing patient referrals.
- Supply chain management.
- Surgical or procedural kitting.
- Benchmarking and quality reporting.
- Cardiovascular information systems.

Medtronic took big financial and operational risks with this endeavor, but not without rewards! On an average, efficiency savings range between 20–25% at partner hospitals. Patient throughput times and waiting lists have decreased. Physician and nurse satisfaction have improved, and patients are happier with their experience.

Meanwhile, Medtronic had also quietly built up $500 million in contracts to manage cath labs in hospitals in several European nations.[23] It is scaling the cath lab program and expanding its services to include post-therapy, home care, and monitoring. "Our Hospital Solutions group is clearly a business model innovation incubator that is now growing to become a business unit delivering these services," said Rob Ten Hoedt, senior Vice President and President, Europe, the Middle East, Africa, and Canada. "We have developed the model ourselves and we are scaling up rapidly, also outside of Western Europe".[24]

Overall, according to Mr. Ishrak, Medtronic has now nearly 40 long-term agreements with hospital systems, including 21 agreements added

---

[22] PwC case studies. http://www.pwc.com/us/en/health-industries/medical-technology-innovation/case-studies.jhtml. Leading examples in medtech innovation practices — Case Study: Medtronic's Cath Lab Program — Sharing risk for treating cardiovascular disease.
[23] http://www.startribune.com/business/272883721.html.
[24] PwC case studies. *Op. Cit.*

> this quarter from their NGC acquisition, which in total represents over $900 million of committed revenue over an average contract period of five to six years. In the previous quarter, Ishrak put this number at half a million, so while small, this business definitely has demand.[25] And that isn't surprising. Hospitals are desperately looking to become more efficient and any company that can help hospitals achieve this will be a winner in the future.

More and more, we are moving toward a people-centered focus for innovation. Companies are investing in ways to get closer to their customers, to gain a better understanding of the motivations and needs of the people who will use their products and services through social media, living labs and design thinking, amongst others. Rather than the traditional transactional relationship between the buyer and the seller, relationships with the customer today need to be more creative and interactive. In addition, managing well the cycle of products and services over time can generate long-term, ongoing value to the customer and the company as well.

CORPORATE SUSTAINABILITY STRATEGIES

The fast pace of the innovation-driven economy has left many companies disorientated within the maze of a diverse and rapidly shifting competitive environment. At the same time, companies are increasingly called upon to take significant responsibilities for resolving some of the world's most intractable problems; from climate change, to community impacts of business operations, and the health of employees.

In the race to conform to the latest sustainability trend, companies are losing sight of the bigger picture. They need to redefine their strategies to cope with the new reality. In this section, we will discuss a few corporate sustainability strategies through which companies can connect sustainability with business approaches in order to develop integrative and long-term solutions to environmental and social challenges.

Circular Economy

Today's economy relies heavily on extracting raw materials from the earth and processing these into parts and products, which are eventually discarded. As the raw materials and the energy required for production

[25] Medical devices business (2014). *Op. Cit.*

become more and more scarce and costly, an alternative approach to the traditional 'linear' economy is required.

The circular economy aims to decouple economic growth from the use of natural resources and ecosystems. A far more effective use of materials results in greater value creation. The approach revolves around innovation in materials and reuse of components and products through disruptive technology and business models based on longevity, renewability, reuse, repair, upgrade, refurbishment, capacity sharing, and dematerialization. Companies no longer focus mainly on driving more volume and squeezing out cost through greater efficiency in supply chains, factories, and operations. Rather, they concentrate on rethinking about products and services from the bottom up to "future proof" their operations to prepare for inevitable resource constraints, all the way through to the customer value proposition. This implies eliminating waste, creating step changes in resource productivity and at the same time enhancing the customer value proposition on dimensions such as price, quality, and availability.

Circular economy principles can be applied to product design, business models, reverse cycle processes, and/or other building blocks:

- Circular design, i.e., improvements in materials selection and product design (standardization/modularization of components, purer material flows, and design for easier disassembly), lies at the heart of a circular economy.
- Innovative business models, especially changing from ownership to 'pay for performance' models, are instrumental in translating products designed for reuse into attractive value propositions.
- Core competencies along reverse cycles involve establishing cost-effective, better-quality collection and treatment systems (either by manufacturers themselves or by third parties).
- Enablers for improving cross-cycle and cross-sector performance are factors that support the required changes at a systems level and include higher transparency for material flows, alignment of incentives, and the establishment of industry standards for better cross-chain and cross-sector collaboration. Other aspects are access to financing and risk management tools, regulation and infrastructure development, and last but not least education, to increase general customer awareness and to create the skill base to drive circular innovation.

The circular economy drives new business models, solutions and services, shifting from "transactions" to "relationships" *via* service and solution models. Products are designed for disassembly and serviceability. Furthermore, the conventional model of customer ownership of the product is replaced by customer access to the product, which is, "paying for performance".

Organizational Zoom ...

Philips's Transition from Linear to Circular Economy

"In Philips, we are fundamentally redesigning our business and end-to-end value chains".

— Frans van Houten
CEO of Royal Philips

In June 2013, Philips became a strategic partner of the Ellen McArthur foundation, a leading organization on the concept of Circular Economy.[26] Following the partnership, a two-year project plan on Circular Economy was established to embed the concept in the relevant businesses, building from the existing knowledge of the Philips Healthcare refurbishment business and the recycling activities from Lighting and Consumer Lifestyle. From an economic point of view, embedding the Circular Economy concept promises additional value creation potential across Philips value chains.

Philips has a track record in the collection and recycling of lamps. For example, in the EU, Philips has a stake in 22 collection and service organizations that collect 40% of all mercury-containing lamps put on the market, with a recycling rate greater than 95%.[27] Recently, Philips started to also sell lighting as a service based on its 'pay per lux' lighting model, which allows customers to use the best light solution for their needs without owing the hardware. At the end of the contract period, Philips lighting products can be taken back into their production process again and raw materials be reused. It's a new way for customers to achieve their sustainability goals: High lighting performance, high energy efficiency, and a low materials footprint.

At Philips Healthcare, sustainability is an integrated part of Refurbished System. Philips had started the refurbishment of medical imaging

[26]http://www.philips.com/philips/sites/philipsglobal/about/sustainability/ourenvironmentalapproach/greeninnovation/circulareconomy.page.

[27]http://reports.weforum.org/toward-the-circular-economy-accelerating-the-scale-up-across-global-supply-chains/how-it-works-up-close-case-examples-of-circular-products/#view/fn-19.

systems in 1989. Their most recent move is a new healthcare imaging systems refurbishment facility inaugurated in Best, the Netherlands this year.[28] Under their refurbishment programs, pre-owned equipment is converted into 'as-new system' with full warranty and reconfigured customers' needs by using the renewed components, latest software release and updates. With reuse in mind, the design and development approach will be enhanced to re-enter a maximum amount of pre-owned product elements into the production process once their first life cycle ends, being a key element of the circular economy.

Resource restriction means that materials reuse is the most economically and environmentally reliable model for future corporate success. Exploring the reuse, remanufacture and refurbishing of materials within the value chain can lead to significant savings for businesses, whilst also generating less dependence on the need for resource input from raw materials. In addition, by applying circular approaches and thinking across their value chains, companies can transform their products and services by rethinking the way they do business, which in turn can identify new revenue streams, product or service innovations and new co-creations. The McKinsey group, assigned by the Ellen McArthur foundation,[29] the leading organization on the concept of Circular Economy, produced two reports on the value creation possibilities of a Circular Economy. For the durable products and solutions industry, these reports indicate an economic potential of EUR 250–500 billion per annum for Europe alone in the next decade.

Moreover, companies that become more resource efficient promote themselves as responsible businesses that care about the markets they serve. Companies that go further by thinking about circular strategy also help create wider positive impacts beyond the value chain. It is these companies, who work to live within the constraints of the resources, which will ultimately develop more competitive, cutting-edge business models and lead to a more restorative economy.

Reverse Innovation

For most of the last two centuries, innovations have flowed one way, from rich countries to poor countries — global companies develop high-end

---

[28] http://www.theguardian.com/sustainable-business/gallery/2014/nov/25/integrating-circular-systems-into-healthcare-in-pictures.
[29] http://www.ellenmacarthurfoundation.org/.

products and adapt them for emerging markets. However, in the past few years, relationships between developed and developing economies became a two-way street, with innovation, investment and competition originating in both the developed and emerging markets, and moving across the world from one to the other end. Indeed, we see increasing innovations flow in the opposite direction from poor to rich economies. Especially, the term "Reverse Innovation" was first coined by Vijay Govindarajan to describe this unconventional trend of innovation. Essentially, it refers broadly to the process whereby goods developed as inexpensive models to meet the needs of emerging (or developing) countries, such as battery-operated medical instruments in countries with limited infrastructure, are then repackaged as low-cost innovative goods in developed markets.

Reverse innovation is driven by the income gap between mature and emerging markets. A product designed for the American mass market is difficult to be simply adapted to meet the needs of the Chinese or Indian mass market. As Vijay Govindarajan explains, "Buyers in poor countries demand solutions on an entirely different price-performance curve. They demand new, high-tech solutions that deliver ultra-low cost and 'good enough' quality".[30]

International corporations have predominantly followed a simple global strategy for a long time. First, innovations happened at home, and then the new offerings would be exported with at most some minor adaptations to address local market needs. This strategy worked well in the past, as the biggest markets were other rich countries.

As already discussed in a previous section, it is not the case anymore. In the last decade, emerging markets have evolved into innovation hubs for new products. They also represent a rapidly growing middle class, a good source of potential customers, and of educated and motivated talent. International corporations have started to recognize that research and development are increasingly important in developing markets. Now they are entering a new stage — to develop market-oriented products in and for emerging economies through globalized innovation teams, which are meant to be scaled up for worldwide use from the beginning.[31] Hence the classical value

[30] http://www.tuck.dartmouth.edu/people/vg/blog-archive/2009/10/what_is_reverse_innovation.htm.

[31] J Immelt, V Govindarajan and C Trimble (2009) How GE is disrupting itself. *Harvard Business Review* 10, pp. 56–65.

creation approach of Western MNCs turns upside down, with a very high competitive level of future competition.

Organizational Zoom . . .

Reverse Innovation in Leading MNCs[32]

GE India — Steam Turbines

In 2010, GE's Indian arm tied up with Triveni Engineering and Industries Ltd. to manufacture steam turbines in the 30–100 MW range. The partnership was to take advantage of lower input costs incurred in manufacturing in India and export these products to West Asia, Indonesia, Europe and the Latin America.

Coca-Cola — eKOCool

Coca-Cola's Indian arm Hindustan Coca-Cola Beverages introduced eKOCool, a chest cooler operating on solar energy with a capacity to store about four dozen 300 ml glass bottles. The innovation also charges mobile and solar lanterns. Coca-Cola has plans to pilot the innovation in different cities in India and may be it will introduce it in other developed countries as well.

LG — Low-cost Air Conditioners (ACs)

South Korea based LG Electronics (LG) plans to develop low-cost ACs targeting the middle and lower-middle classes in India. Their goal is to manufacture ACs at the cost of air coolers which are very common in India.

Wal-Mart — Small format stores in Mexico

Wal-Mart learnt a lesson in Mexico. Mexican shoppers prefer smaller stores compared to the large format stores which Wal-Mart owned in the US. By 2012, Wal-Mart owned 1,250 small stores (Bodegas Aurrera stores) out of the total 2,138 stores in Mexico. Wal-Mart then opened similar small-format stores in the US and Latin America.

Reverse innovation have the potential to help a company achieve sustainability. Reverse innovation is associated with designing product in a simple yet functional way. By scaling down the product and restraining functionalities to the core, resources required in manufacturing the product (e.g., material, human, and logistics, etc.) are significantly reduced.

[32]*Source*: http://www.casestudyinc.com/reverse-innovation-definition-and-examples.

Yet, these kinds of innovations are not only about redesigning existing products, but also rethinking the whole product development process from the scratch, especially for MNCs.[33] Indeed, companies realize that the incremental cost reductions achieved by simplifying products can reach a plateau, and they actually take radical steps of going back to the drawing board and designing afresh in and for emerging markets.

Stakeholder Engagement

Today, the resources and capabilities available to companies for product and service development are more widely shared, but they are accordingly complex to navigate and leverage successfully. The possibilities that exist for leveraging innovation today are immense. Entire networks of users, suppliers, and technology lie at the doorstep of the enterprise waiting to be utilized. The trend toward cross-sector collaboration and stakeholder dialogues is clear. And the underlying idea is that partnerships between businesses and other sectors can drive sustainable development.

Some leading companies are tapping into these trends and understand the interdependence between societal and company performance on the global scale. Stakeholder engagement provides opportunities to further align business practices with societal needs and expectations, helping to drive long-term sustainability and shareholder value. Inability to sustain good relationships with stakeholders can have a damaging effect, not just on reputation but also on actual project results.

In this increasingly complex and ever-changing business environment; engaging different stakeholders is a valid mechanism for promoting sustainable innovation and development for firms. Not only do companies have to look within the walls of the organization to see where creativity and innovation can be leveraged, they also have to look externally to understand their environmental and social impacts and concerns, and look at the entire ecosystem and value chain to transform the resources of the ecosystem into valuable innovations to address the sustainability concerns.

[33]The Economist (2010) First break all the rules, 395(8678), pp. 6–8. http://www.economist.com/node/15879359.

Organizational Zoom...

Xerox's Stakeholder Engagement Practices[34]

Focus	Engagement Practices	Examples
Customers	• Sentinel Customer Satisfaction® Measurement System: Real-time customer feedback in a closed-loop process. • Annual Customer Satisfaction Measurement Survey. • Competitive Benchmark Surveys.	• In 2007, Xerox launched their Sustainability Forum to create awareness of the importance of sustainability, share best practices and encourage customers to reduce their environmental footprint by finding "Smarter Ways to Green." • Every day, a company officer is assigned to respond to all customer concerns coming into headquarters that day by following through to resolution. • Exchange is an online interactive site to engage in dialogue with our customers on issues that affect their business.
Employees	• Employee Engagement Survey conducted regularly. • Sentinel employee feedback tool on Intranet.	• Senior executives participate in bimonthly interactive Web Chats broadcast live over the company's Intranet to employees around the world.

[34]Xerox company website. http://www.xerox.com/Static_HTML/citizenshipreport/2008/stakeholder-page1-1.html.

Focus	Engagement Practices	Examples
	• Ethics hotline. • Open-door policy. • Global Webcasts. • Town hall meetings. • Employee round-tables.	• CEO Anne Mulcahy hosted 40 employee round-tables or town hall meetings, and President Ursula Burns hosted 17 employee round-tables with employees from around the world. • The Sentinel Customer Satisfaction Assurance System™, Xerox proprietary Web-based polling/routing/reporting system, links employee inquiries or comments through the Intranet with appropriate Xerox contacts to identify and track any comment, suggestion or unresolved problem for action.
Shareholders	• Annual meeting of shareholders. • Quarterly teleconferences. • One-on-one investor briefings. • Annual investor conference.	• The company hosted 23 small group meetings with potential investors in major markets including Boston, Chicago, New York, San Francisco, Los Angeles, Dallas, and Kansas City. • Xerox executives spoke with investors/analysts through more than 150 one-on-one meetings and phone calls.

Focus	Engagement Practices	Examples
Suppliers	• Quarterly business reviews with key suppliers. • Review key suppliers' compliance with the EICC code of conduct.	• 50% of compliance reviews were completed with key suppliers in high-risk regions; the remaining reviews are planned for completion by year-end.
Communities	• Employees contributed their time and talents in hundreds of community-related projects worldwide.	• More than 70,000 students have participated in the US-based Xerox Science Consultants Program since its start 40 years ago. • In Poland, through the "Xerox — Better Start!" program, students team up with Xerox employees to learn first-hand about working at a large global company. The students, who are orphans and/or physically handicapped, gain on-the-job experience from their Xerox mentors and leave the program with the skills necessary to begin their careers.
Policy makers & influencers	• Through active memberships with business and trade associations and frequent engagements with public policy makers, Xerox helps to educate on key issues related to its business industry and stakeholders.	• Xerox is partnering with the Rochester Institute of Technology's Golisano Institute for Sustainability to provide programs that embrace the principles of sustainability in product development.

Focus	Engagement Practices	Examples
		• Xerox joins Microsoft, Intel, IBM and other high-tech leaders as a founding corporate member of the Information Overload Research Group. IORG is dedicated to boosting workplace productivity by fighting information overload through the development of solutions to sort and categorize the flood of electronic information from cell phones, e-mail, instant messages, and billions of printed pages worldwide.

Internally, a firm's ability to produce new products and services is inextricably linked to its capability to organize its engagement with its main internal stakeholder, the employee. Companies can promote innovation through human resource management (HRM) practices, such as recruitment and selection, training and development, feedback, reward and recognition of employees. For example, during the recruitment process, organizations can use relatively sophisticated practices such as psychometric tests, assessment centres and work sampling activities to select the people who have the knowledge and skills required to experiment with new ideas.[35] Organizations should also provide on-job education and training to further enhance employees' knowledge and skills that are critical to learning and innovation. In the context of the sustainable enterprise, one important purpose of human resource system is to achieve a strong commitment and participation from all of the company members to innovate toward

[35]H Shipton, D Fay, M West, M Patterson and K Birdi (2005) Managing people to promote innovation. *Creativity and Innovation Management* 14(2), pp. 118–128.

sustainability.[36] Hence, companies engaging with employees through HRM practices which are aligned with the sustainability strategy are contributing to fostering employee creativity in order to find new products, solutions, or processes that substantially improve the existing ones.

A company's external stakeholders include customers, suppliers, community, and non-government organizations. Engaging with and responding to external stakeholders help companies establish credibility and support for their license to operate. It is especially critical for multinational companies to capture inputs of stakeholders in specific markets to understand local impacts. Regular dialogues with local community members not only help organizations to meet compliance requirements and secure permits for specific location, but also prove invaluable as a way of working through incidents or conflicts. Moreover, companies are now engaging stakeholders to get ahead in addressing emerging issues. By focusing on those issues that are most important to stakeholders, materiality analysis better equips a company with insights that can foster innovation, including the development of new business approaches, products, and services. Investors, for their part, are beginning to recognize that companies which routinely engage stakeholders on sustainability issues are also typically leaders in risk management and innovation.

INNOVATING TOWARD A SUSTAINABLE FUTURE: CONCLUDING REMARKS

Environmental issues have become increasingly serious over the last couple of decades. Companies are facing significant challenges and opportunities from environmental issues such as soil degradation, water acidification, air pollution, waste generation, and carbon emissions, to name a few. If the excessive and unsound use of natural resources continues and nothing is done to address the situation, the total demand for limited resources stocks (like biomass, fossil energy, and many metals) is expected to reach 130 billion tons by 2050. That is up from 50 billion in 2014 and it will result in more than 400% overuse of Earth's total capacity: A feat that's physically impossible.[37] Even with a relatively optimistic forecast for technological innovation and improvements in resource efficiency, demand for limited

[36] JE Ricart, MA Rodríguez, P Sánchez and L Ventoso (2005) *The Sustainable Enterprise: Learning from DJSI Leaders.* Bilbao: Fundación BBVA.

[37] Accenture report (2014) Circular advantage: Innovative business models and technologies to create value in a world without limits to growth. http://www.accenture.com/

resources will be 80 billion tons by 2050. The economic impact of resource scarcity on this scale would be devastating.

Stakeholders in these businesses including governments, shareholders, business customers, local communities, consumers, employees, and NGOs are demanding that companies become more focused on sustainability for themselves and the communities they impact.

To solve these problems, we need to reduce the volumes of physical goods that are manufactured and consumed to an adequate, manageable size, and yet without decreasing our current quality of life. Hence, we must shift to a new paradigm that aims at qualitative satisfaction rather than quantitative sufficiency. In other words, economic growth should be decoupled from consumption of materials and energy.[38]

To achieve this paradigm, products should have higher value and should be based primarily on knowledge and service content, rather than mere materialistic worth.[39] In this context, the concept of 'Service Innovation' has been attracting considerable attention, as it creates new value by integrating products and services. Services create value without abandoning materials, and therefore can continuously create value throughout the entire life cycle of a product. Values that would otherwise be lost through wasted materials are instead maintained or even improved by repairing, upgrading, remanufacturing or remarketing products. And additional revenue is generated; thanks to extended usage.

Importantly, the service model provides an alternative to the traditional "buy and own" model. Products are used by one or many customers through a lease or pay-for-performance arrangement. This business model turns incentives for product durability and upgradability upside down, shifting them from volume to performance. It also promotes a platform for collaboration among product users, either individuals or organizations. These facilitate the sharing of overcapacity or underutilization, increasing productivity and user value creation. With a service model, product longevity, reusability, and sharing are now drivers of revenues and reduced costs.

SiteCollectionDocuments/PDF/Accenture-Circular-Advantage-Innovative-Business-Models-Technologies-Value-Growth.pdf.

[38]T Tomiyama (1997) A manufacturing paradigm toward the 21st century. *Integrated Computer Aided Engineering* 4(1997), pp. 159–178.

[39]SL Vargo and RF Lusch (2004) Evolving to a new dominant logic for marketing. *Journal of Marketing* 68(1), pp. 1–17.

Consequently, service innovation can be an approach for dematerializing products without losing value.[40] In this dematerialization approach by services, the level of customer satisfaction generated toward a service plays an important role in compensating for the compromise in materialistic value of the products. For example, in the literature of environmentally conscious design, it has been pointed out that environmentally conscious products need to decrease environmental impacts in consideration of customer demands.[41]

In the face of pressing resource challenges and rising customer and policy expectations for better, more sustainable products, there's never been a better time to start. By developing a proactive strategy that is built on a clear understanding of the motivation to leave behind the old paradigm and underpinned by the business models, technologies, and capabilities critical to success, companies can get ahead of rivals by innovating for both resource efficiency and customer value.

---

[40] E Sundin and B Bras (2005) Making functional sales environmentally and economically beneficial through product remanufacturing. *Journal of Cleaner Production* 13(9), pp. 913–925.

[41] T Sakao and Y Shimomura (2007) Service engineering: A novel engineering discipline for producers to increase value combining service and product. *Journal of Cleaner Production* 15(6), pp. 590–604.

BIBLIOGRAPHY

Abercrombie, S. (1984). *Architecture as Art.* New York: Harper & Row.

Adler, P. and Winograd, T. (eds.) (1992). *Usability: Turning Technologies into Tools.* New York: Oxford University Press.

Andersen, B., Howells, J., Hull, R., Miles, I. and Roberts, J. (eds.) (2000). *Knowledge and Innovation in the New Service Economy.* Aldershot, UK: Edward Edgar.

Anthony, S.D. (2012). *The Little Black Book of Innovation.* Boston: Harvard Business Press.

Anthony, S.D. *et al.* (2008). *The Innovator's Guide to Growth; Putting Disruptive Innovation to Work.* Boston: Harvard Business Press.

Bandura, A. (1986). *Social Foundations of Thought and Action: A Social Cognitive Theory.* Englewood Cliffs, NJ: Prentice-Hall.

Belliveau, P., Griffin, A. and Somermeyer, S. (2002). *The PDMA Toolbook for New Product Development.* New York: Wiley and Sons.

Bettencourt, L. (2010). *Service Innovation: How to Go from Customer Needs to Breakthrough Services.* New York: McGraw Hill.

Berkun, S. (2010). *The Myths of Innovation.* Sebastopol: O'Reilly Media.

Boucharenc, C. (2008). *Design for a Contemporary World.* Singapore: NUS Press.

Brown, T. (2009). *Change by Design: How Design Thinking Transforms Organizations and Inspires Innovation.* New York: Harper Collins.

Chesbrough, H.W. (2003). *Open Innovation: The New Imperative for Creating and Profiting from Technology.* Boston: Harvard Business School Press.

Chesbrough, H.W. (2011). *Open Service Innovation, Rethinking your Business to Grow and Compete in a New Era.* Hoboken, New Jersey: Jossey-Bass-Wiley.

Christensen, C.M. (1997). *The Innovator's Dilemma.* Boston: Harvard Business School Press.

Christensen, C.M. and Eyring H.J. (2011). *The Innovative University.* San Francisco: Jossey-Bass.

Christensen, C.M., Horn, M.B. and Johnson, C.W. (2008). *Disrupting Class: How Disruptive Innovation Will Change the Way the World Learns.* New York: McGrawHill.

Das, T.K. (ed.) (2014). *Strategic Alliances for Innovation and R&D.* Charlotte: Information Age Publishing.

D'Aveni, R.A. (1994). *Hypercompetition: Managing the Dynamics of Strategic Maneuvering.* New York: Free Press.

Debra, M.A. (1997). *Innovation Strategy for the Knowledge Economy: The Ken Awakening.* UK: Butterworth–Heinemann.

Drucker, P. (1974). *Management: Tasks, Responsibilities, Practices.* New York: Harper & Row.

Duffy, F. and Henney, A. (1989). *The Changing City.* London: Bulstrode Press.

Fairbank, J. K. (1987). *The Great Chinese Revolution 1800–1985.* New York: Harper Perennial.

Faktor, S. (2012). *Econovation, The Red, White, and Blue Pill for Arousing Innovation.* Hoboken, New Jersey: John Wiley & Sons.

Florida, R. (2002). *The Rise of the Creative Class: And How it's Transforming Work, Leisure, Community and Everyday Life.* New York: Perseus Book Group.

Gallouj, F. and Djellal, F. (2010). *The Handbook of Innovation and Services, A Multi-disciplinary Perspective.* Cheltenham: Edward Elgar.

Gruen, V. (1964). *The Heart of Our Cities: The Urban Crisis, Diagnosis and Cure.* New York: Simon & Schuster.

Haour, G. (2004). *Resolving The Innovation Paradox.* London: Palgrave Macmillan.

Heskett, J.L. (1986). *Managing in the Service Economy.* Boston: Harvard Business School Press.

Heskett, J.L. (2012). *The Culture Cycle.* Upper Saddle River, New Jersey: FT Press.

Heskett, J.L., Sasser, W.E. and Schlesinger, L.A. (1997). *The Service Profit Chain.* New York: Free Press.

Hillier, B. (1996). *Space is the Machine.* Cambridge: Cambridge University Press.

Jackson, J.B. (1994). *A Sense of Place, a Sense of Time.* New Haven: Yale University Press.

Kao, J. (2007). *Innovation Nation, How America Is Losing Its Innovation Edge.* New York: Free Press.

Keenan, B. (1977). *The Dewey Experiment in China: Educational Reform and Political Power in the Early Republic.* Cambridge: Harvard University Press.

Kelley, T. and Littman, J. (2005). *The Ten Faces of Innovation: IDEO's Strategy for Defeating the Devil's Advocate and Driving Creativity Throughout Your Organization.* New York: Currency Doubleday.

Komninos, N. (2008). *Intelligent Cities and Globalisation of Innovation Networks.* Abingdon, UK: Routledge.

Lee, Ch.M., Miller, W.F., Hancock, M.G. and Rowen, H.S. (eds.) (2000). *The Silicon Valley Edge. A Habitat for Innovation and Entrepreneurship.* Stanford, CA: Stanford University Press.

Leonard-Barton, D. (1995). *Wellsprings of Knowledge.* Boston: Harvard Business School Press.

Lockwood, T. (ed.) (2010). *Design Thinking, Integrating Innovation, Customer Experience, and Brand Value.* New York: Allworth Press.

Lynch, K. (1960). *The Image of the City.* Cambridge: The MIT Press.

Manu, A. (2007). *The Imagination Challenge.* Berkeley: New Riders.

March, J.G. (1994). *A Primer on Decision Making: How Decisions Happen.* New York: The Free Press.

Markman A.B. and Wood K.L. (eds.) (2009). *Tools for Innovation.* Oxford: Oxford University Press.

Martin, R.L. (2009). *The Design of Business: Why Design Thinking is the Next Competitive Advantage.* Boston: Harvard Business School Press.

Mathe, H. and Pavie, X. (2014). L'innovation, élan du XXIe siècle. Paris: L'Harmattan.

Mathe, H. and Shapiro, R.S. (1993). *Integrating Service Strategy in the Manufacturing Company.* London: Chapman & Hall.

Menzies, G. (2002). *1421 — The Year China Discovered the World.* London: Bantam Books.

Midgley, D. (2009). *The Innovation Manual.* Sussex, UK: Wiley.

Mitchell, W.J. (1999). *E-topia.* Cambridge: MIT Press.

Moon, Y. (2010). *Different, Escaping the Competitive Herd.* New York: Crown Business.

Neumeier, M. (2009). *The Designful Company.* Berkeley: New Riders.

Norman, D.A. (2004). *Emotional Design.* New York: Basic Books.

OECD (2014). *OECD Science, Technology and Industry Outlook 2014.* Paris: OECD Publishing.

Olins, W. (1989). *Corporate Identity: Making Business Strategy Visible Through Design.* London: Thames & Hudson.

Osterwalder, A. and Pigneur, Y. (2010). *Business Model Generation.* New York: Wiley.

Pavie, X., Scholten, V. and Carthy, D. (2014). *Responsible Innovation, From Concept to Practice.* Singapore: World Scientific.

Pine, B.J. and Gilmore, J.H. (1999). *The Experience Economy.* Boston: Harvard Business School Press.

Prahalad, C.K. and Krishnan, M.S. (2008). *The New Age of Innovation.* New York: McGraw Hill.

Prahalad, C.K. and Ramaswamy, V. (2004). *The Future of Competition: Co-creating Unique Value with Customers.* Boston, MA: Harvard Business School Press.

Report of the World Commission on Environment and Development. (1987). *Our Common Future.* Oxford, UK: Oxford University Press.

Rifkin, J. (2014). *The Zero Marginal Cost Society: The Internet of Things, The Collaborative Commons, and the Eclipse of Capitalism.* New York: Palgrave Macmillan.

Rybczynski, W. (1996). *City Life.* New York: Touchstone, Simon & Schuster.

Schein, E.H. (2005). *Organizational Culture and Leadership, 3rd Ed.* San Francisco: Jossey-Bass.

Schroeder, J.E. (2003). Building brands: Architectural expression in the electronic age. In Scott, L.M. and Batra R. (eds.). *Persuasive Imagery: A Consumer Response Perspective.* Mahwah, New Jersey: Lawrence Erlbaum Associates, pp. 349–383.

Sinha, K. (2008). *China's Creative Imperative, How Creativity is Transforming Society and Business in China.* Singapore: Wiley.

Stickdorn, M. and Schneider, J. (eds.) (2011). *This is Service Design Thinking.* Amsterdam, Netherlands: BIS Publishers.

Stiglitz, J.E., Sen, A. and Fitoussi, J-P. (2009). *Report by the commission on the measurement of economic performance and social progress.* CMEPSP. September, 14, 2009. www.stiglitz-sen-fitoussio.fr. In *French*: *Richesse des nations et bien-être des individus.* Paris: Odile Jacob.

Tajani, A., Hahn. J. (eds) (2012). *The Smart Guide to Service Innovation Guidebook Series.* European Commission. Brussels, 2012.

Tan, Y. (2011). *Chinnovation, How Chinese Innovators are Changing the World.* Singapore: John Wiley & Sons (Asia).

Terwiesch, C. and Ulrich, K.T. (2009). *Innovation Tournaments.* Boston: Harvard Business School Press.

Tucker, R. (2002). *Driving Growth through Innovation: How Leading Firms are Transforming their Futures.* San Francisco, CA: Berrett-Koehler Publishing.

Ulrich, K.T. and Eppinger, S.D. (1995). *Product Design and Development.* New York: McGraw-Hill.

Urban, G.L. and Hauser, J.R. (1980). *Design and Marketing of New Products.* Upper Saddle River, New Jersey: Prentice Hall.

Utterback, J., Vedin, B-A., Alvarez, E., Ekman, S., Sanderson, S.W., Tether, B. and Verganti, R. (2006). *Design-Inspired Innovation.* Singapore: World Scientific.

Van Stamm, B. (2003). *The Innovation Wave.* Chichester, UK: John Wiley & Sons.

Verganti, R. (2009). *Design Driven Innovation: Changing the Rules of Competition by Radically Innovating What Things Mean.* Boston: Harvard Business School Press.

Walton, T. (1988). *Architecture and the Corporation. The Creative Intersection.* New York: MacMillan.

Willard, B. (2012). *The New Sustainability Advantage: Seven Business Case Benefits of a Triple Bottom Line.* Canada: New Society Publishers.

Zeithaml, V.A. and Bitner, M.J. (1996). *Services Marketing.* New York: McGraw Hill.

OTHER BOOKS BY THE AUTHOR

Mathe, H. and Pavie, X (2014). *L'innovation, élan du XXIe siècle: Innover Dans l'économie de la Connaissance et des Services.* Paris: L'Harmattan.

Mathe, H. and Tixier, D. (2014). *La Logistique.* Paris: Presses Universitaires de France, Collection "Que sais-je?". New actualized and reedited versions of *La Logistique.* Eight editions: 1985, 1987, 1991, 1996, 1997, 2000, 2005, 2010 with translations into English, Monitor Publisher, Fifth revised edition (2000), Russian (five editions since 1998), Chec (1998), among others.

Mathe, H., Abrate, L., Pavie, X. and Paranthoen, S. (2013). *L'influence du Dialogue sur les Relations et l'Expérience Client.* Paris: L'Harmattan.

Mathe, H., Pavie, X. and O'Keeffe, M. (2012). *Valuing People to Create Value: An Innovative Approach to Leveraging Motivation at Work.* Singapore: World Scientific.

Mathe, H., (ed.) (2008). *L'innovation Dans les Services: Perspectives et Stratégies.* Paris: L'Harmattan.

Mathe, H., (ed.) (2006). *L'innovation, au Coeur de la Competition Dans les Services.* Cergy Pontoise: ESSEC-ISIS.

Mathe, H. (1997). *Le Service Global: Innovations et Stratégies Internationales de Développement Dans les Services.* Paris: Maxima – PUF.

Mathe, H., Tixier, D. and Colin, J. (1996). *La Logistique d'Entreprise: Vers un Management Plus Compétitif.* (1996). Paris: Dunod. Actualized and reedited version of *La Logistique au Service de l'Entreprise.* Four editions : 1983, 1985, 1988 and1993.

Mathe, H. and Shapiro, R. D (1993). *Integrating Service Strategy in the Manufacturing Company.* London and New York: Chapman & Hall.

Mathe, H. and Sparks, D. (1993). *Strategic Marketing for Service Industries.* Zürich and London: Strategic Direction Publisher.

Mathe, H. (ed.) (1991). *Managing Services Across Borders.* Paris: Essec-Eurolog Research Series and the Society of Logistics Engineers.
Mathe, H. (1990). *Service Mix Strategies, a Three-step Guide.* Paris: Essec-Eurolog Research Series and the Society of Logistics Engineers.
Mathe, H. (1989). *Le Service Après-Vente.* Paris: Presses Universitaires de France, Collection 'QueSais-je?' (translated into Russian).

INDEX

'SIA information can be obtained at www.ICGtesting.com
ted in the USA
'W06*1622210915

588BV00002B/5/P

9 789814 719575